Student Guide

The Guide to Surviving General Chemistry

SECOND EDITION

Michael R. Rosen

Editor in Chief
Mark Oram, Ph.D.

Cover Design
Emery Coopersmith

Special Thanks to
Laurie Collins

CENGAGE Learning

Australia • Brazil • Mexico • Singapore • United Kingdom • United States

ISBN: 978-1-305-39138-3

Cengage Learning
20 Channel Center Street
Boston, MA 02210
USA

Cengage Learning is a leading provider of customized learning solutions with office locations around the globe, including Singapore, the United Kingdom, Australia, Mexico, Brazil, and Japan. Locate your local office at: **www.cengage.com/global**.

Cengage Learning products are represented in Canada by Nelson Education, Ltd.

To learn more about Cengage Learning Solutions, visit **www.cengage.com**.

Purchase any of our products at your local college store or at our preferred online store **www.cengagebrain.com**.

Printed in the United States of America
Print Number: 01 Print Year: 2014

TESTIMONIALS

"I really liked the way this book broke down all the topics into such an organized and easy to follow fashion. It took all the useless extra information that I constantly came across in the course textbook and told me what I needed to know."

- Amy Segall

"My problem with chemistry is that I didn't understand the terms my professor used. This book really helped me understand the language, and from that point everything else just fell into place"

- Melanie Rider

"Personally, what I liked best about The Guide to Surviving General Chemistry was the chapter on dimensional analysis. I constantly heard my professor use that term for the first 3 weeks of class and had NO IDEA what he was talking about or why he was even doing it. After Michael explained exactly what it is, why it is important to use, and how to actually do it step by step, it cleared up so much for me! All I needed was those few simple sentences in this book and it just made everything for that whole first month of general chemistry make so much sense. WHY AREN'T THERE OTHER BOOKS LIKE THIS OUT THERE FOR ALL CLASSES?"

-Michael Diamond

"My favorite part of the book is the way the book was designed to feel like it is a one-on-one tutor. The book seems as if I have a tutor on call that I can turn to at any time for a question. I like how I don't have to worry about sifting through information just to find what I am looking for. The book breaks down the information for me just like a tutor would, and tells me what I need to know, or more importantly where to even begin. "

-Matthew Schwartz

"I really like how the book is organized. Everything is alphabetical order and then within each chapter there are subtopics. Just by looking at the table of contents I know exactly where to look, I get a brief run down of what I need to know, and then a sample question with the steps on how to get to the answer. Everything is just so easy to follow. "

- Stacey Rendelman

PREFACE

The Guide to Surviving General Chemistry is one of the only general chemistry books written by a student. Michael Rosen wrote the book as a junior in college after recognizing how many of his peers were struggling with their chemistry courses. Michael has had extensive experience with general chemistry not only after doing well in his own chemistry classes, but by being a chemistry teaching assistant, peer-led-team-learning workshop leader, group tutor facilitator, and private tutor. Being a student himself, Michael knows how frustrating difficult subjects like chemistry can be, and how hard it is to understand the dense vocabulary seen in science textbooks.

Michael set out to write a book that covers all topics seen in general chemistry in a straight forward and effective way that makes sense to students. Through his experience helping students with chemistry throughout his entire college career, Michael knows exactly where students get confused and the typical mistakes that are made on exams. The book consists of his own tips, techniques, and tricks for mastering general chemistry along with example questions with worked out solutions for every topic.

What makes the second edition of this book so unique is that it has been edited by a college chemistry professor, Dr. Mark Oram, who has reviewed all of the information in this book. This book now combines the legitimacy of a conventional textbook with the one-of-a-kind perspective of a student who understands his audience and how not everyone is good at science. Bridging the gap between a professor's level of understanding and that of a student, this book is finally a chemistry text that you can actually understand!

How To Use This Book

This book should be used to supplement the material that you learn in class. For each topic that you discuss in class, find that topic in the table of contents of this book, which is conveniently organized in alphabetical order. Read the entire chapter to first get a feel for what you will be expected to know. Then, go back and read the chapter a second time while paying close attention to the tips and techniques illustrated. By the time the exam rolls around, you should be able to solve the example problems without looking at the step by step instructions.

For those students who wait until the last minute to study for an exam this book can accommodate you as well. If you feel the need to cram right before an exam, you should pay most of your attention to the sentences that follow the asterisks. These sentences highlight the most crucial details that you should know, and the typical tricks that other students fall for in each major topic. Again, it is strongly recommended that you read the entire chapter, but if time is winding down you can read the most important parts which are conveniently marked for you.

Disclaimer

Although this book can be used on its own, it is recommended that it be used to supplement the course textbook and class notes. Also, the author apologizes for any typographical errors.

Contact Information

Please feel free to contact Michael Rosen with any comments, suggestions, or book order inquiries at **SurviveChem@gmail.com** or visit our website at **www.Survivechem.com.**

Table of Contents

CHAPTER 1

ACIDS AND BASES

OVERVIEW

The chemistry of acids and bases often scares introductory chemistry students because it seems like there is a lot of information to know. It might be true that there is a lot that you will be responsible for. However, organizing the information the right way in your head will make the entire subject much easier. This chapter breaks down acid/base chemistry in a very easy and digestible manner. In this chapter we will discuss how to recognize the types of questions that you will be asked; the ways to go about solving them, and tips and strategies that you should keep in mind when approaching these problems.

1. STRONG ACIDS AND BASES

There are 6 strong acids and 6 strong bases: anything else is weak. This means if the acid or base is not one of the 6 strong acids or bases then it is weak. The following table lists the 6 strong acids and the 6 strong bases:

6 Strong Acids		6 Strong Bases	
HCl	HNO_3	$LiOH$	$RbOH$
HBr	H_2SO_4	$NaOH$	$CsOH$
HI	$HClO_4$	KOH	$Ba(OH)_2$

***Easy ways to remember the 6 strong acids and 6 strong bases:

The 6 strong bases are all in a row on the periodic table and form an L shape. If you start at Li and go down to Cs and then one over to Ba, all of the elements in that L shape are the 6 strong bases when bonded to an OH. Therefore there is nothing that you have to memorize because the periodic table will be available on the test.

Out of the 6 strong acids, 3 of them are all in a row on the periodic table. If you look at the halogens in group 17 (or 7A), Cl, Br, and I are all in a row and make up half of the strong acids when bonded to an H. The only compounds that you are going to have to memorize out of the 12 strong acids and bases are HNO_3, H_2SO_4 and $HClO_4$.

Strong acids or bases dissociate completely in water. This means that the concentration of the acid is equal to the concentration of H^+ and the concentration of the base is equal to the concentration of OH^-.

The way that we do this is by using the following equation that accounts for how much the weak acid or base dissociates. For a weak species the equation is:

$$K = \frac{x^2}{[\text{concentration}]}$$

where:x = $[H^+]$

[concentration] is that of the weak acid (or base)

2a. Weak Acids

The typical question that you will be asked for with weak acids is to find the pH, given a K_a value and the concentration of the weak acid. You do this by using the above equation to solve for *x*, which is the H^+ concentration. Once you have this you , you just have to –log that value and to get your pH. Again, the reason why we do all this is because $[H^+]$ is not equal to the concentration of the acid since weak acids do not completely dissociate.

Example: Calculate the pH of 1.0×10^{-4} M HF ($K_a = 1.0 \times 10^{-8}$)

Step 1: Recognize that HF is a weak acid since a K_a value is given and also because HF is not one of the 6 strong acids

Step 2: Write the weak acid equation:

$$Ka = \frac{x^2}{[\text{concentration}]}$$

Step 3: Plug in information and solve for x:

$$1.0 \times 10^{-8} = \frac{x^2}{[1.0 \times 10^{-4}]}$$

$$x^2 = 1.0 \times 10^{-12}$$

$$x = \sqrt{1.0 \times 10^{-12}}$$

$$x = [H^+] = 1.0 \times 10^{-6} M$$

Step 4: -log concentration of H^+

pH= $-\log(1.0 \times 10^{-6})$

Step 5: Solve

pH = 6

**Note that HF is actually a weak acid. A lot of students get this confused because they know that the other halogens are strong acids so they tend to think HF is also strong. For this reason instructors love asking questions using HF because students will often confuse it for a strong acid!

2b. Weak Bases

As soon as you recognize that you are dealing with a weak base you should automatically write down the equation for weak species. One important thing to note about the equation is that when dealing with a weak base you use a K_b value (b for base). Also, the x in the equation is now the concentration of OH^-. This means that for bases not only do you have to figure out $[OH^-]$ and then –log the value, but you will also have to subtract from 14 if being asked for the pH.

Example: Calculate the pH of 1.0×10^{-4} M FrOH ($K_b = 1.0 \times 10^{-7}$)

Step 1: Recognize that FrOH is a weak base since there is a K_b value given and also because FrOH is not one of the 6 strong bases

Step 2: Write the weak base equation:

$$Kb = \frac{x^2}{[\text{concentration}]}$$

Step 3: Plug in information and solve for x:

$$1.0 \times 10^{-4} = \frac{x^2}{[1.0 \times 10^{-7}]}$$

$$x^2 = 1.0 \times 10^{-11}$$

$$x = \sqrt{1.0 \times 10^{-11}}$$

$$x = [OH^-] = 3.16 \times 10^{-6} M$$

Step 4: -log concentration of OH^-:

pOH= -log(3.16×10^{-6})

Step 5: Solve for pOH:

pOH = 5.5

Step 6: Subtract the pOH from 14 to get the pH:

pH = 14 - 5.5 = 8.5

**Don't forget to subtract from 14 because we are dealing with a base

2c. An instructor's curveball

Since there are three different variables in the weak acid/base equation, there are 3 different types of questions that an instructor can ask. In the previous examples we discussed the type of question when you are asked to calculate the pH. These are usually the typical questions you will see on a test.

However, there is one question that teachers love asking and it usually throws a lot of students off. The tricky question about weak acid/bases is one that gives you a K value and a pH. So now you go to your weak acid/base equation and check off the variables that you were given, you will notice that pH is not part of the equation. The trick is, by being given the pH you are actually being given the x in disguise. All you have to do is calculate $[H^+]$, which is the x, from the pH.

The way to calculate $[H^+]$ from the pH is by working backwards. Instead of starting with the H^+ concentration and negative logging it, we need to antilog the negative of the pH to obtain $[H^+]$. In other words we need to calculate 10^{-pH}. To do this, enter the negative of the given pH on your calculator. Next press the second function button and then the log button. You will now be left with $[H^+]$, which you can use to solve the weak species equation as before.

Example: Calculate the concentration of Acetic acid ($K_a = 1.0 \times 10^{-2}$) at pH = 6.

Step 1: Recognize that Acetic acid is a weak acid since you are given a Ka and that it is not one of the 6 strong acids

Step 2: Recognize that you can get to $[H^+]$ from the given pH:

$10^{-pH} = 10^{-6} = x$

Step 3: Plug information into weak acid equation:

$$1.0\times10^{-2} = \frac{(10^{-6})^2}{[\text{concentration}]}$$

Step 4: Solve for Concentration

$$\text{Concentration} = \frac{(10^{-6})^2}{1.0 \times 10^{-2}}$$

Concentration = 1.0×10^{-10} M

3. DILUTIONS AND TITRATIONS

3a. Dilution

A good analogy for dilutions is to think about making a pitcher of iced tea from the packets of crystal light. When you pour a packet of crystal light into the pitcher of water you notice that the iced tea tastes too strong and sugary. In order to fix this problem you simply add more water to the pitcher to make it not taste as strong. By adding water to the pitcher what you are really doing is diluting the iced tea. The reason this works is that adding water increases the volume but decreases the concentration.

The way that we calculate the new concentration after we increase the volume is by using the following formula: $M_1V_1 = M_2V_2$. In this formula M stands for Molarity and V for Volume. Dilution problems typically involve an initial concentration and volume, and being asked to calculate the new concentration or a new volume, once the dilution occurs.

Example: What volume must 55 mL of 1.0 M NaOH be raised to, to get a solution that is 0.025 M?

Step 1: Write the dilution equation: $M_1V_1 = M_2V_2$

Step 2: Identify all the variables that are given in the problem:
M_1= 1.0 M
V_1 = 55 mL
M_2 = 0.025 M
V_2 = ?

Step 3: Plug the values into the equation and solve:
1.0 M × 55 mL = 0.025 M × V_2
So V_2 = (1.0 M × 55 mL)/0.025 M = 2200 mL

3b. Titration

A titration is very common way to determine the concentration of a sample of an acid (or less often a base) by neutralizing the sample with a base of known concentration.

Neutralization occurs once the equivalence point has been reached. The equivalence point is when there is an equal amount of moles of acid and moles of base in the beaker. Usually the way that we know the equivalence point is reached is by using an indicator such as phenophalein, which is colorless in acidic solution but turns pink once the solution turns basic.

Typically what happens in a titration is that we have a known volume of the unknown acid in a beaker, with a few drops of phenophalein added. We then slowly add the base of known concentration from a burette – a tall glass graduated tube that shows what volume of base has been delivered – until the solution just turns pink. At this point neutralization has occurred, and we will then know the volumes of the acid and base used, as well as the initial concentration of the base. From all that we can then determine the acid concentration.

The best way to do this is to calculate the moles of base used from knowing the volume and concentration used. Then, with stoichiometry, we can figure out the moles of the unknown acid. Finally, knowing that and the acid volume we can calculate the concentration in moles/Liter. All of these steps are explained in other chapters in this book. This approach has the advantage that it will work with a titration of ANY acid with ANY base.

However, there is one short-cut that we can use whenever we have a monoprotic acid reacting with a monobasic base. This means that the acid has only one H in its formula, and the base has only one OH in its chemical formula. So HCl and HBr are monoprotic acids, and LiOH and NaOH are monobasic bases. However, H_2SO_4 and $Ca(OH)_2$ would not fit this category, since they each yield more than one H^+ (or OH^-) based on the formula.

The shortcut we use is to use a formula very much like the one discussed above. The only difference is that instead of dealing with the same solution on both sides of the equation we are dealing with an acid on one side and a base on the other. The formula for a titration is $M_aV_a = M_bV_b$, where the subscript 'a' stands for acid and the subscript 'b' stands for base.

Example: When 25 mL of HCl was titrated with 0.25 M NaOH, 35.7 mL of NaOH was needed to reach the equivalence point. What is the concentration of HCl?

Step 1: Recognize that HCl and NaOH are monoprotic or monobasic species, so we can use the dilution equation: $M_aV_a = M_bV_b$:

Step 2: Identify all the variables that are given in the problem:

M_a= ?

V_a = 25 mL

M_b = 0.25 M

V_b = 35.7 mL

Step 3: Plug the values into the equation and solve:
? × 25 mL = 0.25 M × 35.7 mL
So M_a = (0.25 M × 35.7 mL)/25 mL = 0.357 M

4. NEUTRALIZATION REACTIONS

Generic Equation: Acid + Base → Salt + Water

A neutralization reaction is a reaction between an acid and a base. When an acid and a base are mixed, a salt and water are the products. Just like a double displacement reaction, the salt is formed from the negative ion of the acid and the positive ion of the base. The water is formed from the H^+ of the acid and the OH^- of the base to form HOH, or H_2O, which is water.

The typical question(s) that you will be asked in the topic of acid/base chemistry involve a neutralization reaction in which you will need to find the pH of the resulting solution after mixing. A typical question will give you a concentration and a volume of an acid; or a concentration and volume of base. The amount of acid and base present will determine the type of question it is and the way that you are going to have to handle it. There are three different possibilities:

Scenario 1: There is a salt formed but some of the strong acid or base is left after neutralization

Approach: Handle this by taking the -log of the strong species.

Scenario 2: There is a weak acid/base and its conjugate salt left after the neutralization

Approach: This scenario creates a buffer and you must use the Henderson-Hasselbach equation.

Scenario 3: There is an equal amount of acid and base, resulting in complete neutralization of each, and just a salt is left.

Approach: This is a salt hydrolysis question and you must determine whether or not hydrolysis occurred. If it did, you then use the weak/acid base equation.

The remainder of this section illustrates and explains each of these three scenarios much more thoroughly.

Scenario 1: Strong species

After using the dilution equation for both the acid and the base, you notice that there is more of one of the species than the other. In this case there will be more of the first species and the other species will have been completely neutralized and used up.

Whenever there is something strong in excess, it will always dictate the pH of the resulting solution. It doesn't matter what kind of salt is present, it doesn't matter how much water is present: all that matters is the resulting concentration of the remaining strong species. If it is a strong acid, simply –log the concentration and that is your pH. If it is a strong base, -log the concentration and then subtract it from 14 and that is your pH.

Example: What is the pH of the resulting solution when 25 mL of 1 M HF ($K_a = 1.0 \times 10^{-4}$) is mixed with 75 mL of 4 M KOH?

Step 1: Write the equation for the chemical reaction

$HF + KOH \rightarrow KF + H_2O$

Step 2: Write dilution equation

$M_1V_1 = M_2V_2$

Step 3: Perform dilution for the acid to find its concentration after mixing:

$(1\ M)(25\ mL) = (M_2)(100\ mL)$ (Since the volume is the sum of 25 + 75 mL!)

$M_2 = [HF] = 0.25\ M$

Step 4: Perform dilution for the base to find its concentration after mixing:

$(4\ M)(75\ mL) = (M_2)(100\ mL)$

$M_2 = [KOH] = 3\ M$

Step 5: Write the concentrations under the respective compounds in the reaction:

HF +	KOH -->	KF +	H_2O
0.25	3	0	0

Step 6: The <u>smaller</u> starting amount shows the species that gets completely used up and is neutralized by the one in excess. Also, there can only be as much product as the smaller amount of the two reactants (just like with limiting reagents!)

Therefore, subtract the smaller amount from both reactants and take the smaller amount as the amount of product that is made:

HF +	KOH -->	KF +	H_2O
0.25	3	0	0
-0.25	-0.25	0.25	0.25
0	2.75	0.25	0.25

Step 7: Notice that 2.75 M of the strong base is left in solution. Whenever there is something strong left, it will dictate the pH of the final solution. So now you calculate the pH of a 2.75 M solution of KOH:

pOH = – log (2.75) = -0.44

pH = 14 – (–0.44) = 14.44

Scenario 2: Buffers

A buffer is a solution when there is a weak acid and its conjugate salt, or when there is a weak base and its conjugate salt. Whenever there is something weak that is left over in solution you should train yourself to automatically check to see if it is a buffer. Remember, a buffer is a solution that resists changes in pH.

When you identify that a buffer is present, you must use an equation called the Henderson-Hasselbach equation. To find the pH of the buffer solution simply plug in the given concentrations to this equation, along with the pKa, and you can then solve for pH.

The Henderson-Hasselbach equation is:

$$pH = pKa + \log\frac{[salt]}{[acid]}$$

Where: [salt] is the concentration of the salt of the conjugate acid

[acid is the concentration of the weak acid

And pKa is simply the –log(Ka) (Anytime you see 'p' use –log)

Example: What is the pH of the resulting solution when 50mL of 3 M HF ($Ka = 1.0 \times 10^{-4}$) is mixed with 15 mL of 2 M KOH?

Step 1: Write the equation for the chemical reaction

$HF + KOH \rightarrow KF + H_2O$

Step 2: Write dilution equation

$M_1V_1 = M_2V_2$

Step 3: Perform dilution for the acid to find its concentration after mixing:

$(3\ M)(50\ mL) = (M_2)(50\ mL+15\ mL)$
$M_2 = [HF] = 2.3\ M$

Step 4: Perform dilution for the base to find its concentration after mixing:

$(2\ M)(15mL) = (M_2)(50\ mL+15\ mL)$
$M_2 = [KOH] = 0.46\ M$

Step 5: Write the concentrations under the respective compounds in the reaction:

HF +	KOH -->	KF +	H_2O
2.3	0.46	0	0

Step 6: The smaller starting amount shows the species that gets completely used up and is neutralized by the one in excess. Also, there can only be as much product as the smaller amount of the two reactants. Therefore, subtract the smaller amount from both reactants and take the smaller amount as the amount of product that is made:

HF +	KOH -->	KF +	H_2O
2.3	0.46	0	0
-0.46	-0.46	0.46	0.46
1.84	0	0.46	0.46

Step 7: Notice that there is a weak acid (HF) and its conjugate salt (KF) left in solution; so you need to recognize this as forming a buffer. As soon as you recognize that it is a buffer, write down the Henderson Hasselbach equation:

$$pH = pKa + \log\frac{[salt]}{[acid]}$$

Step 8: Calculate pKa by finding the –log of the Ka:

$pKa = -\log(1.0 \times 10^{-4}) = 4$

Step 9: Plug in all of the rest of the values from the bottom row of the table in step 6, and solve. Remember the bottom row is all of the concentrations for the resulting solution after the mixing occurred:

$$pH = 4 + \log\frac{[0.46]}{[1.84]}$$

pH = 3.40

Scenario 3: Salt Hydrolysis

Salt Hydrolysis is a situation that occurs at the equivalence point. When there is an exactly equal amount of acid and base, they both get completely neutralized, resulting in a solution with just a salt left. In other words, if you complete the table of concentrations as in step 6 of the previous two examples and you notice that both acid and base are reduced down to a concentration of zero, then salt hydrolysis occurs.

When there is just a salt left in the resulting solution, we must calculate the pH using the salt. To do this, you must break the salt down into its two ions. For each ion, there are then three things that you have to determine. First, which parent acid or base did each of the ions come from? The positive ion must have paired with the OH^-, so it came from the base. The negative ion must have paired with H^+, so it came from the acid. Second, you want to ask yourself was either of the parents strong or weak? Third, you want to ask yourself whether or not hydrolysis occurs. No hydrolysis occurs if both the parent acid and base are strong. When this happens, the pH of the resulting solution is automatically 7 and there is no work that you need to show. However, if one or both of the parents are weak then hydrolysis occurs.

Refer to the following T chart as an example, using NaF, of how you should treat every salt:

Na	F
NaOH	HF
Strong	Weak
No Hyd	Yes Hyd

In a situation with two strong parents, you will have a salt such as NaCl, and the pH of the resulting solution is simply 7. Write the number 7, circle it, and put a smiley face because you are done. Unfortunately, you will probably not get that lucky on the test. Instead, you will most likely have a situation in which one parent is weak and the other parent is strong. As we have seen, this means hydrolysis occurs. So, whenever you have an answer of yes in the last row of the T chart, you will need to use the equation for weak acids/bases. This makes sense because hydrolysis occurs for weak acids/bases, therefore we are using the weak acid/base equation. Remember the equation is: $K = \frac{x^2}{[\text{concentration}]}$

One thing that I need to point out to you is how to figure out which K value to use. As you know for an acid we use Ka and for a base we use Kb. However, as we just saw, a salt comes from both an acid and a base. The question is then which K value do we use?

The answer is that you use the K value of the strong parent. Therefore, in this case it was a strong base so we will be using a Kb value. If they give you a Ka value in the question, make sure you convert it to a Kb value. The way to convert between Ka and Kb is to use the following relationship: $Ka \times Kb = 1 \times 10^{-14}$
So, divide 1×10^{-14} by the given Ka value to obtain the Kb value first.

After determining the correct K value to use, simply plug in the information into the equation and solve the equation as we did in the weak acid/base section of this chapter. Remember, if you are dealing with a Kb value, this will give you a value for $[OH^-]$, so don't forget to subtract the resulting pOH from 14 if you are asked for a pH!

Example: What is the pH of the resulting solution when 5 mL of 2 M HF($Ka = 1.0 \times 10^{-4}$) is mixed with 10 mL of 1 M KOH?

Step 1: Write the equation for the chemical reaction

$HF + KOH \rightarrow KF + H_2O$

Step 2: Write dilution equation

$M_1V_1 = M_2V_2$

Step 3: Perform dilution for the acid to find its concentration after mixing:

$(2\text{ M})(5\text{ mL}) = (M_2)(5\text{ mL}+10\text{ mL})$
$M_2 = [HF] = 0.67\text{ M}$

Step 4: Perform dilution for the base to find its concentration after mixing:

$(1\text{ M})(10\text{ mL}) = (M_2)(5\text{ mL}+10\text{ mL})$
$M_2 = [KOH] = 0.67\text{ M}$

Step 5: Write the concentrations under the respective compounds in the reaction:

HF +	KOH -->	KF +	H2O
0.67	0.67	0	0

Step 6: Recognize that in this case there is the same amount of acid and base. Therefore this amount is the amount of salt that will be left and the acid and base will completely neutralize each other:

HF +	KOH -->	KF +	H2O
0.67	0.67	0	0
-0.67	-0.67	0.67	0.67
0	0	0.67	0.67

Step 7: Recognize that since there is only salt left so this is a salt hydrolysis problem. As soon as you recognize this, draw the T chart for the salt and determine the 3 pieces of information that you need to continue:

K	F
KOH	HF
Strong	Weak
No Hyd	Yes Hyd

Step 8: Since there was an answer of yes for hydrolysis, use the weak/acid base equation. Also notice that the parent is a strong base so you need to convert Ka given in the problem to K_b (using the relationship $K_a \times K_b = 1.0 \times 10^{-14}$)

$(1.0 \times 10^{-4})(K_b) = 1.0 \times 10^{-14}$
$K_b = 1.0 \times 10^{-10}$

Step 9: Plug in concentration of salt and Kb into weak equation and solve for x:

$$Kb = \frac{x^2}{[\text{concentration}]}$$

$$1.0 \times 10^{-10} = \frac{x^2}{0.67}$$

$$x = [OH^-] = 8.19 \times 10^{-6}$$

Step 10: Solve for pH

pOH = $-\log(8.19 \times 10^{-6}) = 5.09$

pH = 14 – 5.09 = 8.91

**Note that when the salt came from a strong base the pH will be slightly basic and when the salt comes from a strong acid the pH will be slightly acidic.

SUMMARY:

Acid/base chemistry seems to present an insane amount of information that you will need to know. Although there might be a lot of things that you will be responsible for, the material itself is not that bad. The way that I broke this chapter down should be the way that you organize this material in your head. Be sure to recognize the difference between strong acids/bases and weak acids/bases. Also make sure you know how to treat each of them.

For neutralization reactions understand that there are 3 different scenarios. For the first half of all the scenarios you treat the problem identically. Always write out the reaction and then use the dilution equation to calculate the new concentration of acid and base after mixing takes place. The concentration of each will then determine which of the 3 scenarios you will be dealing with. Be able to recognize the 3 scenarios and understand how to treat each one.

The hardest part of acid/base chemistry is recognizing which situation you are dealing with in the problem. Once you can recognize it, you will be in good shape to solve these problems.

CHAPTER 2

COLLIGATIVE PROPERTIES

OVERVIEW

Colligative properties are characteristics of solutions that depend on the number of particles in a given volume of solvent. Colligative properties do not depend on the mass of the particles but rather the amount of particles. The four major topics that fall under this category are vapor pressure; boiling point elevation; freezing point depression and osmotic pressure. In this chapter we will discuss each of these four major areas of colligative properties as well as provide specific examples regarding the types of questions that will be asked.

1. RAOULT'S LAW (VAPOR PRESSURE LOWERING)

Raoult's Law states that the vapor pressure of an ideal solution is dependent on the vapor pressure of each chemical component and the mole fraction of each chemical component.

Definition: $P_A = X_A P_A^o$

Components: P_A = Vapor pressure of solution
X_A = Mole fraction of solvent
P_A^o = Vapor pressure of the pure solvent

This formula is often used to calculate the lowering of vapor pressure. One important thing to note about the formula is the presence of the mole fraction X_A. When dealing with the mole fraction of a single solute, you must subtract this value from 1 in order to get the mole fraction of solvent. (You can get points taken off on these kinds of questions if you forget to subtract the mole fraction from 1!)

An example of this issue is illustrated in the following problem using Raoult's Law:

Example: Given that the vapor pressure of water is 30 torr, what is the vapor pressure of a solution made from 15 g of fructose, which has a molar mass of 180 g/mol, when dissolved in 25 g of water?

Step 1: Recognize that this question involves the use of Raoult's Law because it is asking us to solve for the vapor pressure of a solution while providing us with a P_A^o and enough information to calculate the mole fraction.

Step 2: Write the equation for Raoult's Law

$$P_A = X_A P_A^o$$

Step 3: Calculate mole fraction of the solute

$$X_{solute} = \frac{15g\left(\frac{1mol}{180g}\right)}{=15g\left(\frac{1mol}{180g}\right)+25g\left(\frac{1mol}{18g}\right)}$$

$$X_{solute} = \frac{0.0833}{0.0833+1.389}$$

$$X_{solute} = 0.057$$

Step 4: Convert the mole fraction of solute into the mole fraction of solvent, since the mole fraction of solvent is asked for:

$$X_{solvent} = 1 - X_{solute}$$

$$X_{solvent} = 0.943$$

Step 5: Now that you have all the variables needed, plug in the information into the Raoult's law equation to solve for the vapor pressure of the solution

$$P_A = X_A P_A^o$$

$$P_A = (0.943)(30torr)$$

$$P_A = 28.29torr$$

2. BOILING POINT ELEVATION

Boiling point elevation is a formula used to determine the increase in the boiling point temperature after a certain amount of solute is dissolved in the water.

Definition: $\Delta T_b = +iK_b m$

Components: ΔT_b = Change in boiling point temperature
i= van't Hoff factor
K_b = Boiling constant
m= molality of solution

**Important you will be responsible for knowing that the boiling constant K_b for water is equal to 0.512 $\frac{^oC \bullet kg}{mole}$

**Interesting fact: The reason why you are supposed to add salt to your spaghetti water is because the salt increases the boiling point of the water. With the addition of the salt, the water can now reach temperatures higher than the normal boiling point and will therefore cook the spaghetti more efficiently.

There are a few things that you must be aware of when using the boiling point elevation formula. First, you must be aware of the van't Hoff factor. The van't Hoff measures how many different ions the component breaks down into. For example, NaCl is a salt that breaks apart into Na^+ and Cl^- ions, so i is equal to 2. In a case involving any organic or non-ionic compound, such as a sugar (sucrose, fructose etc) i will be equal to 1.

Another thing that you need to be careful about is to understand exactly what it is that you are solving for. ΔT_b is the increase in boiling point temperature. This means after you plug in all of the information into the formula, you will be left with a number like 3.6 for example. This 3.6 is NOT the new boiling point temperature, but rather the amount the boiling point temperature increased. Therefore, after you get your answer you must add it to the original boiling point temperature.

Finally, in most of the questions you will see, you will be dealing with water so make sure you remember that water boils at 100 °C.

Example: What is the boiling point of a solution made from 4 g NaCl dissolved in 15 g of water?

Step 1: Recognize that this is a boiling point elevation question because it is asking for the boiling point of a solution containing salt.

Step 2: Determine the value of i

NaCl breaks apart into Na^+ and Cl^- ions, meaning i = 2

Step 3: Determine the molality (see Chapter on Concentration Units and Conversions for solutions for how to do this)

$$m = \frac{4gNaCl}{15g_{water}}\left(\frac{1mol}{58g}\right)\left(\frac{1000g}{1kg}\right)$$

$$m = 4.6m$$

Step 4: Solve for change in boiling point temperature

$$\Delta T_b = +ik_b m$$
ΔT_b =(2)(0.512)(4.6)
ΔT_b = 4.7 °C

Step 5: Calculate the elevated boiling point

100 °C + 4.7 °C = 104.7 °C

3. FREEZING POINT DEPRESSION

Freezing point depression is a formula used to determine the decrease in the freezing point temperature after a certain amount of solute is added to the water.

Definition: $\Delta T_f = -ik_f m$

Components:
- ΔT_f = Change in freezing point temperature
- i = van't Hoff factor
- K_f = Freezing constant
- m = molality of solution

**Important you will be responsible for knowing that the freezing point constant for water is equal to 1.86 $\frac{^oC \bullet kg}{mole}$

**Interesting fact: The reason why you add rock salt to sidewalks during a snowstorm is because the rock salt lowers the freezing point of the water. By lowering the freezing point, you are making it harder for ice to form and will therefore prevent people from slipping on ice.

You must still be cautious of the same issues as discussed in the boiling point elevation section. Remember that you are solving for the reduction in the freezing point; meaning that after you obtain an answer you must subtract it from the original freezing point to obtain the new freezing point. The normal freezing point of water is 0 °C.

Example: What is the freezing point of a solution made from 4 g NaCl dissolved in 15 g of water?

Step 1: Recognize that this is a freezing point depression question because it is asking for the freezing point of a solution containing salt.

Step 2: Determine the value of i

NaCl breaks apart into Na^+ and Cl^- ions, meaning i = 2

Step 3: Determine the molality (see Chapter on Concentration Units and Conversions for solutions for how to do this)

$$m = \frac{4gNaCl}{15g_{water}}\left(\frac{1mol}{58g}\right)\left(\frac{1000g}{1kg}\right)$$

$$m = 4.6m$$

Step 4: Solve for the change in the freezing point temperature

$$\Delta T_f = -iK_f m$$

ΔT_f = - (2)(1.86)(4.6)

ΔT_f = -17 °C

Step 5: Calculate the freezing point

0 °C -17 °C = -17 °C

4. OSMOTIC PRESSURE

Osmotic pressure refers to the pressure that is exerted by water in a solution. The osmotic pressure is influenced by the Molarity of the solute, the temperature, the gas constant, and the van't Hoff Factor.

Definition: $\Pi = iM_sRT$

Components: Π = osmotic pressure
i= Van't Hoff Factor
M_s = Molarity
R = Gas constant = $0.0821 L.atm.mol^{-1}.K^{-1}$
T = Temperature in Kelvin

**Important: Since the gas constant R is part of this equation; all units must be in atm, L, and K. Be sure to convert any units that are not already in the correct form.

**Easy way of remembering this equation: Think of "I SMART by osmosis".
By viewing the formula with the s subscript in front of the M, you obtain an expression along the lines of iM_sRT

A multiple choice question that you will often see for osmotic pressure usually asks which one of the given choices exerts the greater osmotic pressure. All of the choices will have a different concentration, van't Hoff factor, or temperature. Since the osmotic pressure equation has all of the variables multiplied by each other, the answer to this type of question will simply be the choice with the bigger number.

Example: Which of the following has the greater osmotic pressure?

A) 0.02 M KF at 300 K
B) 0.02 M NaCl at 500 K

You will notice that the concentrations (0.02 M) and the van't Hoff factors (i = 2) are equal in the two choices. Therefore the question comes down to temperature. Since temperature, as well as every other variable, is directly proportional to osmotic pressure, the answer would be the one with the highest temperature. So the answer to this question would be choice B.

Other questions that you will see for osmotic pressure will typically be questions in which you are given 3 out of the 4 variables in the formula and asked to solve for the 4th.

Example: Two beakers are joined by a membrane. In the first beaker there is a 0.5 M solution of KF and in the second beaker there is pure water. What is the osmotic pressure at 25 °C?

Step 1: Recognize that this is an osmotic pressure question because it is asking for a pressure being exerted by water.

Step 2: Write the equation for osmotic pressure

$$\Pi = iM_s RT$$

Step 3: Plug in information into the equation, with all units converted properly first:

$$\Pi = \frac{(2)(0.5mol)(.0821atm \bullet L)(25+273K)}{(mol \bullet K)}$$

Step 4: Solve

Osmotic Pressure = 24.5 atm

SUMMARY:

For Colligative Properties there are four important equations that you will need to know. Raoult's law is an important means of obtaining a vapor pressure of solution given the mole fraction and the vapor pressure of the components in the solution. Boiling point elevation and freezing point depression are two equations that allow us to determine the change in the initial boiling points and freezing points after a solute is added. Osmotic pressure involves calculating the pressure exerted by water at a given temperature and concentration of solute. Remember, Colligative Properties are properties that involve the NUMBER of particles and not the mass of particles in a given volume.

CHAPTER 3

CONCENTRATION UNITS AND CONVERSIONS FOR SOLUTIONS

OVERVIEW

In this chapter we will discuss all of the different concentration units and conversions that you will be responsible for when dealing with solutions. What is meant by this are units such as Molarity, Molality and mole fraction. This chapter will examine the definitions of each unit, explain the best approach to solving for the unit, and then provide an example question similar to one that would be asked on a test, along with a detailed solution.

Before we get started, I would like to point out the way in which you should go about solving these kinds of questions. Any time you are asked to determine a concentration unit of a solution, I highly recommend that you write the definition of the particular unit that is being asked for. The very first thing that you do is to write the definition down on your paper and use it to serve as a road map. Typically, what happens in these questions is that you are given units in a problem that you will need to convert, using dimensional analysis, into the units that are needed. By writing down the definition of the concentration unit you are being asked for, you will know exactly where it is that you need to get to with dimensional analysis, you will be guaranteed a correct answer!

1. MOLARITY

Definition of Molarity: $M = \dfrac{\text{moles solute}}{\text{Liter solution}}$ units: M

Molarity is the most common concentration unit and is the one that is seen the most on tests and quizzes. The word solution means solute + solvent. For example, a salt dissolved in water would be the solute and the water itself would be the solvent. When the two are put together, that is called a solution. So, Molarity is a "part over total" way of measuring a concentration, since it is the solute (part) over the solution (total).

The best way of solving for Molarity is to identify the number of moles the solute that you are given in the question and then place it in the numerator of your fraction. Then take the volume of solution and put that in the denominator of your fraction. By viewing the line in the fraction as meaning "in" you will be in perfect shape to find the Molarity.

Example: Calculate the Molarity of 0.15 g NaCl in100 mL of solution

Step 1: Write the definition of Molarity

$$M = \frac{\text{moles solute}}{\text{Liter solution}}$$

Step 2: Write a fraction with the information given in the question

$$\frac{0.15\text{ g NaCl}}{100\text{ mL solution}}$$

(This should be read as "0.15 g NaCl in 100 mL solution")

Step 3: Using the definition of Molarity as a road map, first convert grams of NaCl in the numerator to moles of NaCl; then convert mL of solution in the denominator to L of solution. This will leave you with your final answer.

$$\frac{0.15\text{ g NaCl}}{100\text{ mL solution}}\left(\frac{1\text{ mol NaCl}}{58\text{ g NaCl}}\right)\left(\frac{1000\text{ mL solution}}{1\text{ L solution}}\right)=\frac{0.026\text{ mol}}{1\text{ L}}=0.026\text{ M}$$

Note that after the cancellation of units using dimensional analysis we are left with moles solute over L solution, which by definition is our Molarity!

2. MOLALITY

Definition of Molality: $m = \frac{\text{moles solute}}{\text{kg solvent}}$ units: m

Molality, similar to Molarity, is another means of measuring the concentration of a solution. Notice the difference between the two units is in the denominator. Molality uses kg of solvent (that is, just the liquid we dissolve in and not the overall solution!) For this reason, Molality is a ratio measure of concentration, between how much solute we have over how much solvent we have.

Students often get tripped up when calculating Molality. The reason is because, unlike all of the other concentration units, Molality has a mass in the denominator, not a volume. The problem that then arises is that usually a volume is given in the problem, so you will first need to convert the given volume to the mass you need to get Molality.

The way that you must do this is to use density. Since density is measured in grams per mL it is the perfect way of converting back and forth between volume and mass. If the density is given in the question, then use that value for your density. If, however, a density is not given, just assume that the solvent is water and use the density 1 g/mL.

Example: Calculate the Molality of 0.67g KCl when dissolved in 150 mL of a solvent with a density of 1.03 g/mL.

Step 1: Write the definition of Molality

$$m = \frac{\text{moles solute}}{\text{kg solvent}}$$

Step 2: Write a fraction with the information given in the question

$$\frac{0.67\ \text{g KCl}}{150\ \text{mL solvent}}$$

Step 3: Using the definition of Molality as a road map, convert grams of KCl in the numerator to moles of KCl, and then convert mL of solution in the denominator to grams using the given density. Lastly, convert grams to kg to leave you with your final answer.

$$\frac{0.67\ \text{g KCl}}{150\ \text{mL solvent}}\left(\frac{1\ \text{mol Cl}}{74\text{g KCl}}\right)\left(\frac{1\text{mL}}{1.03\ \text{g}}\right)\left(\frac{1000\ \text{g}}{1\ \text{kg}}\right) = 0.06\ \text{m}$$

3. WEIGHT/WEIGHT

Definition of weight/weight: $f = \frac{g_{\text{solute}}}{g_{\text{solution}}}$ units: none

Weight/weight (or w/w) is a very important factor when dealing with conversions of concentration units. The way that I look at weight/weight is that it stands at the top of the mountain. From the top of a mountain you are at the highest point and can see a very long distance in every direction. Relating this back to chemistry, once you find the weight/weight you can go in every direction to any other concentration unit. For all of the remaining concentration units that will be discussed in this chapter, the best way to find them is to first get the weight/weight and from there obtain the unit that you need.

So first let's look at how to obtain the weight/weight, or (w/w).

Example:

Calculate the (w/w) of a 2.34 x 10^{-3} M solution of sucrose (molar mass of 342 g/mol).

Step 1: Write the definition of weight/weight

$$f = \frac{g_{solute}}{g_{solution}}$$

Step 2: Write a fraction with the information given in the question, remembering that Molarity can be written as mol solute /L solution:

$$\frac{2.34 \times 10^{-3}\,\text{mol sucrose}}{1\,\text{L solution}}$$

Step 3: Using the definition of (w/w) as a road map, first convert mol of sucrose in the numerator to grams of sucrose. Next convert L of solution in the denominator to grams of solution, using the density. Remember that since the density was not given, we use 1 g/mL. Once you have g sucrose per g solvent you will then have your final answer!

$$\frac{2.34 \times 10^{-3}\,\text{mol}}{1\,\text{L solution}}\left(\frac{342\,\text{g}}{1\,\text{mol}}\right)\left(\frac{1\,\text{L}}{1000\,\text{mL}}\right)\left(\frac{1\,\text{mL}}{1\,\text{g}}\right) = 8.00 \times 10^{-4}$$

*note that the answer can also be written as: $\frac{8.00 \times 10^{-4}\,\text{g sucrose}}{1\,\text{g solution}}$

4. WEIGHT/WEIGHT PERCENT (%)

Definition of weight/weight percent: w/w % = $\frac{g_{solute}}{100\,g_{solution}}$ units: %

Weight/Weight Percent is very simple. All you have to do is first calculate the w/w and then multiply it by 100. The reason why this is while w/w is grams of solute per 1 g solution, w/w % is grams of solute per 100 g solution.

So, after calculating w/w you will have to get the denominator to be 100 times bigger. According to the laws of math, whatever you do to the denominator you have to do to the numerator in order to maintain the value of the number. So once you calculate w/w, you will then need to multiply both numerator and denominator by 100 to get w/w %.

Example: Calculate the w/w % of a 2.34 x 10^{-3} M solution of sucrose (molar mass of 342 g/mol).

Step 1: Once you calculate the w/w you will have 8.00 x 10^{-4} (w/w) (refer to the above example to see how we calculated this value!)

Step 2: Recognize that this can also be written as: $\dfrac{8.00 \text{ x } 10^{-4} \text{ g sucrose}}{1 \text{ g solution}}$

Step 3: In order to get the denominator to have 100 g of solution you must multiply it by 100. But remember that if you multiply the denominator by 100 you must also do this to the numerator to get your final answer.

$$\frac{8.00 \text{ x } 10^{-4} \text{ g sucrose}}{1 \text{ g solution}} \frac{\times 100}{\times 100} = \frac{0.08 \text{ g sucrose}}{100 \text{ g solution}} = 0.08\,(\text{w/w})\,\%$$

5. PARTS PER MILLION (PPM)

Definition of parts per million: ppm = $\dfrac{\text{g solute}}{10^{6} \text{ g solution}}$ units: ppm

Parts per million is a more practical way of expressing very small concentrations and is an extremely useful concentration unit in real life situations. The only difference between ppm and w/w is that we have 10^{6} g in the denominator of ppm. So to get ppm, first find w/w and then multiply both numerator and demominator by 1 million (10^{6}).

Example: Calculate the ppm of a 2.34 x 10^{-3} M solution of sucrose (molar mass of 342 g/mol).

Step 1: Once you calculate the w/w you will have 8.00 x 10^{-4} (w/w) (refer to the above example to see how we calculated this value!)

Step 2: Recognize that this can also be written as: $\dfrac{8.00 \times 10^{-4} \text{ g sucrose}}{1 \text{ g solution}}$

Step 3: To get the denominator to have 10^6 g of solution you must multiply it by 10^6. But remember that if you multiply the denominator by 10^6 you must also do this to the numerator to get your final answer.

$$\frac{8.00\times10^{-4}\,\text{g sucrose}}{1\,\text{g solution}}\frac{\times10^{6}}{\times10^{6}}=\frac{800\,\text{g sucrose}}{10^{6}\,\text{g solution}}=800\text{ ppm}$$

*Important: Note that the final answer can also be written as $\frac{800\,\text{g solute}}{10^{6}\,\text{g solution}}$.

If you decide to write it this way, DO NOT divide this fraction! I have seen a lot of students get points off because they go ahead and divide 800 by 10^6 and that brings them back to where they started. Where they get confused is that this fraction is the definition of ppm and it is supposed to be written over 10^6 g solution.

6. MOLE FRACTION (*X*)

Definition of mole fraction: $X = \frac{\text{mol solute}}{\text{mol solution}}$ units: none

Mole fraction is a ratio of moles of solute to moles of solution. The most important tip I can emphasize for mole fraction is that the denominator has the word solution. Solution, if you recall, is solute + solvent. This is a very important thing when doing mole fraction and can help avoid a lot of the typical mistakes that students usually make when doing mole fraction.

For this reason I would rewrite the definition of mole fraction as:

$$X = \frac{\text{mol}_{\text{solute}}}{\text{mol}_{\text{solute}}+\text{mol}_{\text{solvent}}}$$

Looking at it this way will help lead you to a more systematic way of solving for mole fraction as illustrated in the following example.

Example: Calculate the mole fraction when 2 g NaCl are dissolved in 200 mL water.

Step 1: Write the definition of mole fraction

$$X = \frac{\text{mol}_{\text{solute}}}{\text{mol}_{\text{solute}}+\text{mol}_{\text{solvent}}}$$

Step 2: Plug in the values given to the respective locations in the definition of *X*:

$$X = \frac{2\ g_{solute}}{2\ g_{solute} + 200\ mL_{solvent}}$$

Step 3: Use dimensional analysis to convert all units to moles (using 1 g/mL for density)

$$X = \frac{2\ g\left(\frac{1\ mol}{58\ g}\right)}{2\ g\left(\frac{1\ mol}{58\ g}\right) + 200\ mL\left(\frac{1g}{1\ mL}\right)\left(\frac{1\ mol}{18\ g}\right)}$$

Step 4: Solve

$$X = \frac{0.034\ mol}{0.034\ mol + 11.11\ mol} = 0.0031$$

** Note: a mole fraction value will <u>always</u> be between 0 and 1. So f you get an initial value outside of this range, unfortunately you will have to recheck your work!

7. NORMALITY

Definition of normality: Normality = Molarity x number of charges   Units: N

Normality is a concentration unit that appears all the time on college level tests and for some reason always takes students by surprise. Normality is similar to Molarity, but the twist is that Normality relates to the fact that an ionic compound, or acid or base, will break apart into charged ions when dissolved, and Normality is the concentration not of the intact solute, but of the total charges on from the ions that are produced.

So all you have to do anytime you are asked for normality is to first find the Molarity and then multiply it by the number of charges on either positively charged cations, or the negatively charged anions, from the given solute.

Finally, you should note that Normality is often used in the context of acids or bases, and when this is the case we are interested in the Normality of H^+ ions (for an acid) or of the OH^- ions (for a base).

Please refer to the following examples for more details.

Example #1: Calculate the Normality of 0.05 M HCl

Step 1: Determine the Molarity

0.05 M (given)

Step 2: Determine totals of the charges on the ions:

H = +1
Cl = -1

Therefore, 1 mole of HCl = 1 mole of positive charge

Step 3: Multiply the molarity by moles of charge to determine normality

0.05 M × 1 = 0.05 N

Another way to do this is by dimensional analysis. For this, remember that M is mol/L, and that 1 mol acid = 1 mol charge:

$$\frac{0.05 \text{ mol HCl}}{1\text{L}}\left(\frac{1\text{mol charge}}{1\text{mol HCl}}\right) = 0.05 \text{ N}$$

Example #2: Calculate the Normality of 0.05 M H_2SO_4

Step 1: Determine the Molarity

0.05 M (given)

Step 2: Determine totals of the charges on the ions:

2 × H^+ = +2 (since there are <u>two</u> H^+ ions from H_2SO_4!)
1 × SO_4^{2-} = -2

Therefore, 1 mole of H_2SO_4= 2 moles of (positive) charge

Step 3: Multiply the molarity by moles of charge to determine normality

0.05 M × 2 = 0.10 N

Or by dimensional analysis: $\frac{0.05\,\text{mol}\,H_2SO_4}{1\text{L}}\left(\frac{2\,\text{mol charge}}{1\text{mol}\,H_2SO_4}\right)=0.10\,\text{N}$

Example #3: Calculate the Normality of .05 M H_3PO_4

Step 1: Determine the Molarity

0.05 M (given)

Step 2: Determine totals of the charges on the ions:

3 × H^+ = +3 (since there are <u>three</u> H^+ ions from H_3PO_4!)
1 × PO_4^{2-} = -3

Therefore, 1 mole of H_3PO_4= 3 moles of (positive) charge

Step 3: Multiply the molarity by moles of charge to determine normality

0.05 M × 3 = 0.15 N

Or by dimensional analysis: $\frac{0.05\,\text{mol}\,H_3PO_4}{1\text{L}}\left(\frac{3\,\text{mol charge}}{1\text{mol}\,H_3PO_4}\right)=0.15\,\text{N}$

SUMMARY

Always handle a concentration unit question by first writing the definition. Use the definition as a road map to help you find the units that you need in the numerator and the denominator. Be careful with the dimensional analysis and make sure all of the units cancel in such a way so that you are left with the units that appear in the definition.

Again, I would also like to stress the difference between solution and solvent. Let's think of an example in which I am preparing salt water in a jug. The salt that I am adding to the jug is the solute. The water that I am putting in the jug is the solvent. Together, the salt and the water that are in the jug make up the solution. So the solution constitutes both the solute and the solvent. This is a very important concept and it is important to keep clear in your mind when obtaining concentration units.

CHAPTER 4

DIMENSIONAL ANALYSIS

1. DIMENSIONAL ANALYSIS

Dimensional analysis, which is also known as the factor label method or unit conversion method, is an extremely important tool in the field of chemistry. In a college setting, a lot of professors do not take the time to explain dimensional analysis and it is often assumed that students already know how to use it. Without having a strong and clear understanding of dimensional analysis, as well as how to use it, students often find themselves struggling when attempting to solve a problem involving unit conversions.

The whole point to dimensional analysis is to convert given values into different units of measure. When converting between units it is important to change the units without changing the value of the number. The way that this is accomplished is by multiplying our given value by fractions which are equal to 1. As we know, when a quantity is multiplied by 1, the quantity remains unchanged. By having an equivalent value in the numerator and denominator of a fraction, the overall fraction is equal to 1 and the amount of the original value is preserved. However, the units will change by this approach, which is exactly what we want!

Normally the fractions that we use in dimensional analysis are called conversion factors. Lets first talk about the kids of conversion factors we can use, and then look at how to do some examples of dimensional analysis conversions.

1a. Conversion factors for different units

In general chemistry you typically will not see too many questions that ask for a straight unit conversion. However, it is important to have a general understanding of the conversion factors so that you can use them if you need to convert between units within a much larger chemistry problem.

There are many conversion factors that students often memorize but never usually come up on a test or quiz. The conversion factors that usually appear the most, and that I suggest you commit to memory are the following:

1 ft = 12 in
1 yd = 3 ft
1 mi = 5280 ft
1 in = 2.54 cm
1 atm = 760 torr (we will see this a lot in gas laws chapter)
1 cal = 4.183 J (we will see this a lot in thermochemistry chapter)
1 cm^3 = 1 mL (This one converts between units of volume, and is used very often)
1 kg = 2.21 lb

**Note: It is very important to check with your instructor if a list will be supplied. Some professors do supply tables of conversions, and this will save you time in an exam. So make sure to ask your instructor if they will do this!

1b. Conversion factors for metric units

Other than the conversions that were just previously noted, another set of conversions that are very important in chemistry are conversions within the metric system. An example would be converting from meters to centimeters. The easiest way to approach these is to keep in mind that the metric system of units is based on powers of ten; so all you have to do when converting metric units is simply move the decimal place.

In order to know which direction to move the decimal place, as well as how many places it should be moved, I devised a phonemic device to remember how to convert within the metric system. To use this device, first you need to memorize the following phrase: "King Henry Doesn't Mind discussing church matters". This should help you recall the order of the metric prefixes: namely Kilo, Hecto, Deca, Meter, deci, centi, milli.

To use this in practice, all you have to do is note which prefix you are starting with and which one you need to convert into. Then, for every letter you move towards the prefix you are converting to, that's how many decimal places you move! For instance, if you want to convert from kilometers to meters, you start with you the phonemic device underlining the two prefixes that you are interested in:

K H D M D C M

Next, you count the amount of letters you are passing including the letter that you are stopping with. In this particular example we are moving three places to the right. So start with 1 kilometer then move the decimal point of the 1 three places to the right. This will give you 1000, so this means that 1 kilometer is equal to 1000 meters! This works with any unit conversion within the metric system and becomes particularly important with dimensional analysis.

One last thing to keep in mind about metric conversions is that the smaller number always goes with the bigger unit. Meaning, we know that there is a difference between 3 decimal places between kilo and meter but some people get confused as to which number goes with which unit. The smaller number always goes with the bigger unit (kilo) and the bigger number always goes with the smaller unit (meter). This is why there is 1 kilometer in 1000 meters and not the other way around.

1c. Using conversion factors to convert between units

Suppose we have to convert 95 meters into miles. Let's see how to do this in detail, using dimensional analysis.

When using dimensional analysis, we need to find conversion factors that relate the unit we start with to the one that we want. However, if we can't find one directly, we will need to take multiple steps to accomplish this. Working backwards from the unit that we want to get to is an excellent approach in situations where a path between units is not completely apparent. So, in the above example where we have to convert 95 meters to miles, the first thing you want to ask yourself is what unit, based on the table of conversions given earlier, could get us to miles? The answer to this question would be feet. But in order to get to feet we then see that we would need inches. Then, to get to inches we know we would need centimeters. Since we are starting with meters we could simply first go from meters to centimeters and then follow the path that we just outlined.

If it helps, try writing out the path from one unit to another using backwards arrows:

$$\text{meters} \leftarrow \text{centimeters} \leftarrow \text{inches} \leftarrow \text{feet} \leftarrow \text{miles}$$

Once you identify the path to use between the two units, you next arrange one or more a conversion factors in such a way that each added will cancel the unit that comes before it. In this example, since we are starting with meters, we then want to convert first from meters to centimeters, so we use the conversion factor 1m = 100 cm (remember King Henry?!). To do this, turn 1 m = 100 cm into a fraction, and arrange it so the unit that is given (meters) will cancel, and the next unit (centimeters) will take its place when multiplied together. Remember that when we multiply fractions the numerator in the first will cancel the denominator in the second. The following shows this first step:

$$95\ \cancel{m}\left(\frac{100\ \text{cm}}{1\cancel{m}}\right)$$

Now, we are left with centimeters after meters cancels. Thinking back to our path, we next want to go from centimeters to inches. We can do this using 1 in = 2.54 cm. We will again cancel the existing unit (cm) by putting that unit in the denominator, and the new unit in the numerator. The following is an illustration of the second step:

$$95\ \cancel{m}\left(\frac{100\ \cancel{cm}}{1\ \cancel{m}}\right)\left(\frac{1\ \text{in}}{2.54\ \cancel{cm}}\right)$$

Next, we are left with inches and we need to get to feet. Inches will be put in the denominator to cancel inches in the numerator of the previous step. We then use the conversion factor 12 in = 1 ft to complete the third step:

$$95\ \cancel{m}\left(\frac{100\ \cancel{cm}}{1\ \cancel{m}}\right)\left(\frac{1\ \cancel{in}}{2.54\ \cancel{cm}}\right)\left(\frac{1\ \text{ft}}{12\ \cancel{in}}\right)$$

Lastly, we are left with feet and need to get to miles. Feet will be put in the denominator to cancel feet in the numerator of the previous step. We will then convert to miles where we will be left with our desired unit as shown in the last step:

$$95\ \cancel{m}\left(\frac{100\ \cancel{cm}}{1\ \cancel{m}}\right)\left(\frac{1\ \cancel{in}}{2.54\ \cancel{cm}}\right)\left(\frac{1\ \cancel{ft}}{12\ \cancel{in}}\right)\left(\frac{1\ \text{mi}}{5280\ \cancel{ft}}\right)$$

Now that we are left with the unit that we are looking for, all we need to do is the math. The way to go about calculating the answer is by multiplying all the numbers in the numerators and then dividing that quantity by all the numbers multiplied in the denominators. The way that this should be plugged into the calculator is, (95x100x1x1x1)/(1x2.54x12x5280) and you will get an answer of 0.059 miles:

$$95\ \text{m}\left(\frac{100\ \text{cm}}{1\ \text{m}}\right)\left(\frac{1\ \text{in}}{2.54\ \text{cm}}\right)\left(\frac{1\ \text{ft}}{12\ \text{in}}\right)\left(\frac{1\ \text{mi}}{5280\ \text{ft}}\right)=0.059\ \text{mi}$$

1c: Approaching conversions with multiple units at a time

Once you have mastered dimensional analysis, you will be able to handle even more complicated conversions, such as ones that involve two different units that need to be converted. A typical example of this would be to convert between meters/second to miles/hour. You do this the exact same way; just handle one unit at a time. First take care of converting meters to miles and once that is complete, continue with converting seconds to hours. This is shown in the following example that converts 36m/s to mi/hr:

$$\frac{36\text{m}}{1\text{s}}\left(\frac{100\text{cm}}{1\text{m}}\right)\left(\frac{1\text{in}}{2.54\text{cm}}\right)\left(\frac{1\text{ft}}{12\text{in}}\right)\left(\frac{1\text{mi}}{5280\text{ft}}\right)\left(\frac{60\text{s}}{1\text{min}}\right)\left(\frac{60\text{min}}{1\text{hr}}\right)=\frac{81\text{mi}}{\text{hr}}$$

SUMMARY

As you can see dimensional analysis is not as scary as it appears. It provides for a very straightforward approach to convert between units. A lot of people like to steer clear of dimensional analysis as much as possible, but I highly recommend using it wherever possible. By using dimensional analysis you will have a very neat, organized, and guaranteed approach to solving chemistry questions. Your entire math is right there in one easy step without having it all over the page in multiple step calculations. Also, by using dimensional analysis you will be able to check that all of the units cancel and that the only unit left is the unit in your final answer.

2. SIGNIFICANT FIGURES

'Significant figures' is a term that many students hate hearing. Significant figures also referred to as sig figs, are the very reason why students don't earn full marks on their tests and quizzes even when their answers are otherwise fully correct. You might ask yourself, how can an instructor take off points for a correct answer? Well, if your answer does not contain the correct amount of significant figures then the instructor reserves the right to do so. (Remember however you should check before each test with your instructor exactly what their policies are on this point. It is your right to do so!)

Significant figures address the issue of uncertainty. Each measurement that we make has a certain amount of uncertainty associated with it. The last digit in every number is the one that contains the uncertainty. The total significant figures in a measurement tell us how certain we are of a particular number. The example that I always like to use is the amount of time it takes me to walk to the library. If I tell you that I am a 5 minute walk away from the library that leaves a lot of room for question. Do I really mean that I am slightly less than a 5 minute walk, such as about 4.5 minutes, or do I mean I am slightly above a 5 minute walk/ such as about 5.5 minutes?

In other words, by just saying the number 5, it leaves a lot of room for interpretation. Now, let's say instead of saying 5 minutes I had said 5.00 minutes. By saying 5.00 now we are confident I am pretty much exactly five minutes away, with far less room for uncertainty. And this in tur is because the number 5.00 is a much more precise number than just 5, because it contains three significant figures as opposed to just one (see next section).

Now understanding the correlation between sig figs and uncertainty, we should examine the rules for determining how many sig figs there are in a number. This next section will not only lay out the exact rules that you need to know to determine the sig figs correctly, but will also explain why we use such a system in the first place.

2a: Rules for determining significant figures

When determining if a digit is significant or not, you have to ask yourself whether or not that particular digit leads to a more precise number. Does the digit help improve the certainty of the number, or does it just act as a place holder? The following are the rules that you must use to determine the number of significant figures in a number:

1. All non-zero digits are significant
   Example: 534 contains 3 significant figures

2. Zeros between two non-zero digits are significant
   Example: 103 contains 3 significant figures

3. Place holder zeros are NEVER significant
   Example: 0.0001 contains 1 significant figure

4. Trailing zeros are significant in numbers less than 1
   Example: 0.23000 contains 5 significant figures

5. Trailing zeros are not significant in numbers WITHOUT a decimal point at the end
   Example: 1000 contains 1 significant figure

6. Trailing zeros are significant in numbers if a decimal IS present at the end
   Example: 1000. contains 4 significant figures

When generalizing the rules, the hardest part is remembering which zeros are significant and which are not. Thinking in terms of whether a particular zero contributes to a more precise number will help you in your determination.

When you think about it, the zeros in the number 0.004 do not tell us anything about how confident we are about that number, they simply serve as place holders that tell us that the value is 4 thousandths of what is being measured, as opposed to hundredths or tenths. However, the zeros in .4500 are serving to tell us that we have 4 tenths, 5 hundreds, zero thousandths and zero ten-thousandths: so we are much more confident in the accuracy of this reading In other words, these zeros ARE significant.

The same holds true for the zeros in numbers greater than one. The number 1000 only has 1 significant figure when there is no decimal point at the end of the number. But by putting a decimal at the end we make clear that we are talking about exactly 1000. If the decimal point is not there, the zeros are essentially serving as place holders, stating that the number is measured in the thousands.

2b: Significant figures in calculations

Being able to correctly count sig figs is only half the battle. The other half is determining how many sig figs your answer should have. The rules for this depend on the math operation that you are dealing with. In all math operations, our answer can only be as precise as the number with the least precision in our calculation. In other words, if we are multiplying two quantities together, and one of the numbers had 20 significant figures but the other number only had 2, then our answer can only have two. We can only be as certain as our most uncertain number. An analogy that I like to give for this is that a team can only be as strong as its weakest member.

Adding and Subtracting

In adding and subtracting we focus on the amount of **decimal places** in the numbers we are adding or subtracting. The number with the least amount of decimal places limits the amount of decimal places our answer should have.

Example:

	20.012	← 3 decimal places
+	4.3	← 1 decimal place
	24.3	← 1 decimal place only, since the 2nd value had only 1 decimal place

Multiplying and Dividing

In multiplying and dividing we focus on the total amount of **significant figures** in the numbers we are multiplying and dividing. The number with the least amount of significant figures limits how many significant figures our answer should have.

Example:

	101.40	← 5 significant figures
×	.0034	← 2 significant figures
	0.34	← 2 significant figures, since the 2nd value had only 2 sig figs

Combination of Adding and Subtracting with Multiplying and Dividing

On tests, instructors really like asking questions that involve both rules at the same time. In cases like this, it is important to only do one operation at a time. Do each operation one at a time applying the rules for that specific operation, and keep reducing the expression all the way down to a single number. An example of this is illustrated in the one below:

$$\frac{(12.05+2.9)(11.45-3.132)}{(6.83 \text{ x } 10^2)(6.100)} = \frac{(15.0)(8.32)}{(4170)} = \frac{(125)}{(4170)} = 0.0300$$

As you can see, each operation is handled one at a time being very careful not to intermix the two different rules for the two different operations. The reason for this is because adding and subtracting focuses on decimal places and multiplying and dividing focuses on significant figures. For this reason we cannot just plug the entire expression into the calculator and then just look at each of the original numbers. The reason why our final answer has 3 sig figs is because after following the rules for each operation we were left with a 3 significant figure number divided by a 4 significant figure number.

Another important point to note is the number $6.83 \text{ x } 10^2$. This number is written in what is called scientific notation. The way to handle numbers written like this is to ignore the number 10 and its exponent when assessing significant figures and decimal places. Do not count the number 10 or the digits of the exponent as significant figures.

Rounding and the use of a calculator

Although calculators might very well be the best thing that has ever happened to you, they can sometimes be your worst nightmare if not treated properly. When you type expressions into the calculator, the calculator spits out an answer without thinking. It is your job to know how many digits in read out you should write down as your final answer. For instance, know you need three significant digits in your answer, but your calculator might spit out 8. You must then stop that number at three significant digits by rounding at the last digit.

The normal rounding rules that you have learned back in elementary school apply. If the number to the right is lower than 5 you round down, and greater than or equal to 5 you round up. Please refer to the following example:

186.43 ← 5 significant figures
× 2.10 ← 3 significant figures

Calculator readout: 391.503 Answer has 5 significant figures but needs 3

Rounding: 391|.503 digit to the right is greater than or equal to 5

Final answer: 392

SUMMARY

Remember that significant figures are digits that have a direct impact on increasing the certainty of a number. Final answers can only be as certain as the number with the least amount of certainty in the starting values. Do not trust your calculator read out and be sure to follow the correct rules of rounding to obtain your final answer with the correct significant figures.

CHAPTER 5

ELECTROCHEMISTRY AND REDOX

OVERVIEW

Electrochemistry and redox reactions both revolve around the same general principle relating to the transfer of electrons. This principle is Oxidation and Reduction. In this chapter we will examine what is meant by the terms oxidation and reduction as well as go over the general concepts of both redox and electrochemistry. By the end of this chapter you will understand exactly what will be expected of you for each topic and you will know exactly how to approach the types of questions that you will be asked.

1. REDOX

1a. Oxidation

The term oxidation describes the loss of electrons from a chemical compound. A compound that loses electrons is said to be oxidized. As we know, electrons are negatively charged and therefore compounds that lose electrons become more positively charged. Oxidation is the first half of a redox reaction, and for this reason we can write what is called a 'half reaction' to describe what happens during oxidation. The following is an example of an oxidation half reaction:

$$Zn^{+2} \rightarrow Zn^{+4} + 2\ e^{-}$$

1b. Reduction

The term reduction describes the gain of electrons by a chemical compound. A compound that gains electrons is said to be reduced. The gain of electrons causes a compound that is reduced to become more negatively charged. Reduction is the second half of a redox reaction. When a compound is oxidized, the electrons that are lost by that compound are used to reduce another compound which in turn gets reduced. The following is an example of a reduction half reaction:

$$Mg^{+2} + 2\ e^{-} \rightarrow Mg^{0}$$

1c. Determining Oxidation Numbers

The overall objective of redox is to determine which element is being oxidized and which element is being reduced in the full reaction. In order to determine this, we need to be able to calculate an oxidation number for elements within chemical compounds. When you hear the words 'oxidation number', just think of the word charge. Some elements can have more than one charge or oxidation number. It is our responsibility to determine which oxidation number is present on each element in order to look for oxidation and reduction.

The following are a list of things that you need to keep in mind when determining oxidation numbers:

1. The oxidation number of a single pure element by itself is always 0
2. In a neutral compound, the total positive charges must equal the total negative charges
3. In a compound that has a net charge, the sum of the positive and negative charges should be equal to the net charge
4. When group 1 and group 2 metals are present in a compound, they will always have an oxidation number of +1 and +2 respectively
5. When oxygen is present in a compound, it will almost always have a -2 oxidation number
6. When halogens are present in a binary compound, they will usually have a -1 oxidation number
7. When hydrogen is present it will have a +1 oxidation number in a covalent compounds and a-1 oxidation number in ionic compounds

Example: what is the oxidation number of Mn in MnO_4^-?

$[MnO_4]^-$

+7 -8

Answer: +7

Explanation: $[MnO_4]^-$ is a polyatomic anion with a net charge of -1; so the sum of the oxidation numbers of each of its elements must add to -1. Oxygen always has the oxidation number of -2. Since there are 4 oxygen atoms in the molecule, there is a total charge of -8 from the oxygen. In order to make the overall compound total then be -1, Mn needs to have an oxidation number of +7.

1d. Writing the whole redox equation

We previously talked about the two halves of a redox reaction. We will now put the two halves together, along with our knowledge that we just learned about oxidation numbers and we will determine which element got oxidized and which got reduced.

Example: Which elements get reduced and oxidized in: $2\ Mg + O_2 \rightarrow 2\ MgO$

Step 1: Determine oxidation numbers for every element, using the list above

$2\ Mg^0 + O_2^0 \rightarrow 2\ Mg^{+2}O^{-2}$

Step 2: Determine which element got more positive, or oxidized:

$$2\ Mg^0 + O_2{}^0 \rightarrow 2\ Mg^{+2}O^{-2}$$

(Mg^0 → Mg^{+2}: oxidized)

Step 3: Determine which element got more negative, or reduced:

$$2Mg^0 + O_2{}^0 \rightarrow 2\ Mg^{+2}O^{-2}$$

($O_2{}^0$ → O^{-2}: reduced)

Follow this same procedure every time. First, determine the oxidation numbers of each element. Then look for the element which became more positive as it went from a reactant to a product. In this case it was Mg that got oxidized. Next determine which element became more negative as it went from a reactant to a product. In this case it was O_2 that was reduced.

Aside from knowing which element got oxidized and which element got reduced, you will also encounter another set of important terms. These terms are oxidizing agent and reducing agent.

A **reducing agent** is the element that causes the reduction to occur. In all cases this would be the element that gets oxidized. If you think about it logically, the element that gets oxidized loses electrons which then go on to reduce another element. Therefore, the reducing agent, or the agent responsible for the reduction, is the element that got oxidized. In this case the Mg is the reducing agent.

An **oxidizing agent** is the element that causes the oxidation to occur. By the same argument, this would be the element that gets reduced. The element that gets reduced must be getting electrons from another element, which is the reason why oxidation is taking place. The oxidizing agent in this case is O_2.

1e. Balancing redox half reactions in acid and base

One last thing that you will need to know is how to balance redox equations. Redox equations must be balanced by both mass and charge. In other words, not only do all of the elements on one side of the equation have to be equal to the elements on the other side of the equation, but the charges must be equal as well.

Also, instructors will ask you to balance redox equations in either acid or base conditions. The best way to go about balancing the equations, regardless of whether or not it is done in acid or base conditions, is to first balance all of the non-hydrogen and non-oxygen elements in the equation. This means you need to balance all of the elements except for the hydrogen and oxygen elements. Once you have done that, you then need to balance the oxygen elements using H_2O. You do this by adding H_2O molecules to the side of the equation that has fewer oxygen atoms. Next, you balance hydrogen elements by adding H^+ to the side of the equation with the fewer hydrogen elements. At this point you are now ready to balance the charges. Add up all of the charges on the first side of the equation and compare it with the sum of all the charges on the second side of the equation. Add as many electrons as are needed to the side of the equation that has too much positive charge to balance this out.

When balancing in acidic conditions, you would be finished after following the steps outlined above. However, if you were balancing in basic conditions there is one additional step that you will have to take. In basic conditions, there are no H^+ ions in solution. Therefore, if we used H^+ to balance the hydrogen elements, we cannot leave our final equation with the H+ in it. To get rid of the H^+ in the equation, we must use another equation; the water ionization equation, or $H_2O \rightarrow H^+ + OH^-$. We do this to cancel all of the excess H^+ to leave us with just water and hydroxide ions in our final balanced equation under basic conditions.

The next two examples will clarify these steps for acid or base conditions:

Example 1: Balance $MnO_4^- \rightarrow Mn^{+2}$ in acidic conditions

Step 1: Balance all elements except hydrogen and oxygen

$MnO_4^- \rightarrow Mn^{+2}$

Step 2: Balance oxygen by adding water:

$MnO_4^- \rightarrow Mn^{+2} + 4\ H_2O$

Step 3: Balance hydrogen by adding H^+:

$8\ H^+ + MnO_4^- \rightarrow Mn^{+2} + 4\ H_2O$

Step 4: Balance the charges:

$8\ H^+ + MnO_4^- + 5\ e^- \rightarrow Mn^{+2} + 4\ H_2O$

Since we had to balance in acid conditions, we are finished.

Example 2: Balance $MnO_4^- \rightarrow Mn^{+2}$ in basic conditions

Step 1: Balance all elements except hydrogen and oxygen

$MnO_4^- \rightarrow Mn^{+2}$

Step 2: Balance oxygen by adding water:

$MnO_4^- \rightarrow Mn^{+2} + 4\ H_2O$

Step 3: Balance hydrogen by adding H^+:

$8\ H^+ + MnO_4^- \rightarrow Mn^{+2} + 4\ H_2O$

Step 4: Balance the charges:

$8\ H^+ + MnO_4^- + 5\ e^- \rightarrow Mn^{+2} + 4\ H_2O$

Step 5: Eliminate H^+ by using the water ionization equation:

$8\ H^+ + MnO_4^- + 5e^- \rightarrow Mn^{+2} + 4H_2O$
$8\ H_2O \rightarrow 8\ H^+ + 8\ OH^-$

$MnO_4^- + 8\ H_2O + 5\ e^- \rightarrow Mn^{+2} + 4\ H_2O + 8\ OH^-$

Step 6: Cancel like terms and write the final answer:

$MnO_4^- + 4\ H_2O + 5\ e^- \rightarrow Mn^{+2} + 8\ OH^-$

1f. Writing the complete redox reaction

Once you have two balanced half reactions, you combine them to form a complete redox reaction. But first you must first make sure the coefficient in front of the e^- is the same in both reactions. If not, then you need to multiply the equations in order to make both e^- coefficients the same. After this, add the two equations and cancel out like terms. In other words, compounds that appear on the reactant side of one half reaction and the product side of the other half reaction cancel each other out.

Example:

$2 \times (MnO_4^- + 4\ H_2O + 5\ e^- \rightarrow Mn^{+2} + 8\ OH^-)$
$5 \times (Zn^{+2} \rightarrow Zn^{+4} + 2\ e^-)$

$2\ MnO_4^- + 8\ H_2O + 10\ e^- + 5\ Zn^{+2} \rightarrow 2\ Mn^{+2} + 16\ OH^- + 5\ Zn^{+4} + 10\ e^-$

$2\ MnO_4^- + 8\ H_2O + 5\ Zn^{+2} \rightarrow 2\ Mn^{+2} + 16\ OH^- + 5\ Zn^{+4}$

2. ELECTROCHEMISTRY

Electrochemistry involves the movement of electrons and the processes of oxidation and reduction. In electrochemistry, the most frequent topic studied is that of a voltaic cell. A voltaic cell consists of a spontaneous chemical reaction in which one species gets oxidized and the other gets reduced. (By contrast, an electrolytic cell does not consist of a spontaneous chemical reaction and requires an external power source.)

Before we start talking about voltaic cells, lets first look at some important terms to be familiar with in electrochemistry.

2a. Important Terms

Cathode: The site at which reduction occurs. The cathode is what the electrons flow towards and usually is written on the right side of a voltaic cell.

Anode: The side at which oxidation occurs. The anode is what the electrons flow out from and is usually written on the left side of the voltaic cell.

**Good way of remembering this information: "An Ox-red Cat" (meaning anode-oxidation, reduction – cathode)

2b. Voltaic Cell

A voltaic cell, which is also known as a galvanic cell, is a set-up that you will see a lot in electrochemistry. A voltaic cell consists of two separate half cells, where each half cell has its own half reaction: one for oxidation and the other for reduction.

As you will notice in the diagram below, the half cell on the left is where electrons are flowing out from, and therefore the electrode is being oxidized. Since oxidation occurs at the anode this makes the half cell on the left the anode.

The electrons then flow to the half cell on the right where the electrode there gets reduced. So this half cell is the cathode , since it is the site where reduction occurs.

To complete the loop, the ions produced at the cathode travel through a salt bridge which connects the two half cells and allows for the reaction to continue.

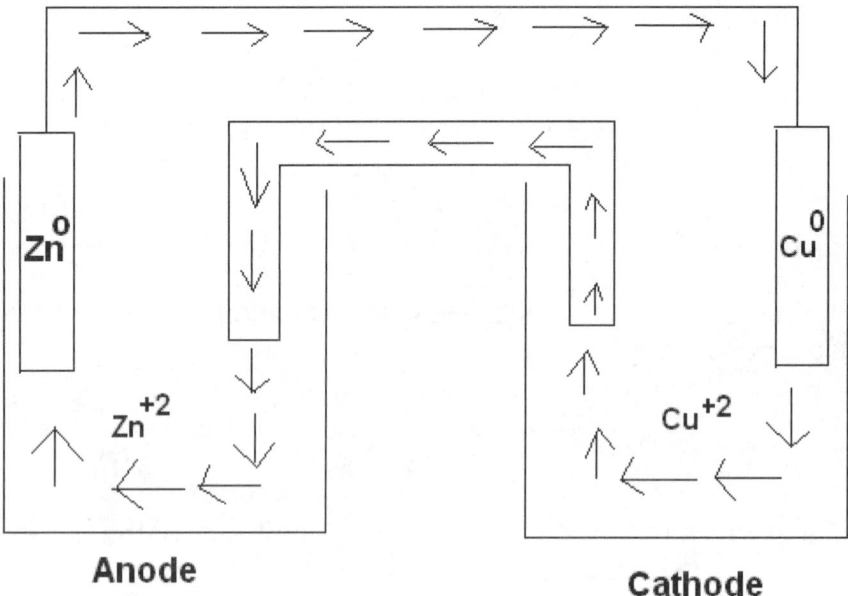

2c. Writing half reactions for electrochemistry

In the voltaic cell above, there are two half reactions that are occurring. The first is the oxidation of Zn and the second is the reduction of $Cu2^{+}$. Writing a half reaction in electrochemistry is the same as writing a half reaction for redox. However, there is one additional piece of information that will be present. Each half reaction will have an standard electrostatic potential associated with it. You will see this written as E^{o}. The standard potentials are all listed on a standard table that you will be provided with.

Examples of standard reduction potentials:

Zn^{+2} +2 e^{-} → Zn^{0} E^{o} = -0.76 V

Cu^{+2} + 2 e^{-} → Cu^{0} E^{o} = +0.34 V

One important thing to note is that all of the standard potentials on the table are for reduction reactions. Therefore, if you need a standard potential for an oxidation reaction you must negate the standard potential as it appears on the table.

With the standard potentials of the two half reactions, you can then calculate the standard potential of the cell. The standard potential of the cell is expressed as $E^{o}_{cell.}$ To find the standard potential of the cell, you use the formula: $E^{o}_{cell} = E^{0}_{ox} - E^{0}_{red}$. This formula means that you subtract the potential of the reduction reaction from the potential of the reduction reaction. A good way of remembering this is "subtract the most negative number from the most positive number."

Example: Calculate the E^o_{cell} for the voltaic cell above

Step 1: We already saw that Zn is oxidized. So first reverse this reaction AND the sign of the standard potential:

$Zn^0 \rightarrow Zn^{+2} + 2\,e^-$ $E^o = +0.76$ V

$Cu^{+2} + 2\,e^- \rightarrow Cu^0$ $E^o = -0.34$ V

Step 2: Write the equation for E^o_{cell}

$$E^o_{cell} = E^0_{ox} - E^0_{red}$$

Step 3: Plug in values

$$E^o_{cell} = +0.76 - (-0.34)$$

Step 4: Solve

$$E^o_{cell} = +1.10\text{ V}$$

** Note: For a voltaic cell, the answer will always be positive. If you get a negative value, check to see you identified the anode and cathode reactions properly!

2d. The Nernst Equation

$$E_{cell} = E^o_{cell} - \frac{RT}{n} \ln \frac{[\text{oxidised species}]}{[\text{reduced species}]}$$

(n = coefficient in front of e^-)

The Nernst equation is an important equation in electrochemistry. An important thing to note in electrochemistry is that there is a difference between E^o_{cell} and E_{cell}. The difference is that E^o_{cell} is the electrostatic standard potential of the cell and E_{cell} is the actual electrostatic potential of a given cell. A good way to think of the Nernst equation is that is provides us with a means of applying the standard potentials from the table to a specific situation at a given temperature and point in the cell reaction. In other words, the Nernst equation gives us a means for us to calculate the E_{cell} from E^o_{cell}.

An easier way of using the Nernst Equation, and the way that you will most likely always be using it, is at the standard temperature 25 °C. Here the Nernst Equation is:

$$E_{cell} = E^o_{cell} - \frac{0.05916}{n} \log \frac{[\text{oxidised species}]}{[\text{reduced species}]}$$

Example: Calculate E_{cell} of a non-standard cell with 1 M Zn^{+2} and a .03 M Cu^{+2} at 25 °C.

Step 1: Write Nernst Equation at 25 °C:

$$E_{cell} = E^{o}_{cell} - \frac{0.05916}{n}\log\frac{[\text{oxidised species}]}{[\text{reduced species}]}$$

Step 2: Calculate E^{o}_{cell} from standard potential table, as we did above:

$$E^{o}_{cell} = E^{0}_{ox} - E^{0}_{red}$$

$$E^{o}_{cell} = +0.76 - (-0.34)$$

$$E^{o}_{cell} = +1.10\ \text{V}$$

Step 3: Determine value for n:

$Zn^{0} \rightarrow Zn^{+2} + 2e^{-}$
$Cu^{+2} + 2e^{-} \rightarrow Cu^{0}$

Both reactions have a coefficient of 2 in front of e^{-} therefore n =2

Step 4: Plug in values to equation:

$$E_{cell} = 1.10 - \frac{.05916}{2}\log\frac{[1.0]}{[0.3]}$$

Step 5: Solve

$$E_{cell} = +1.0\ \text{V}$$

SUMMARY

Redox and electrochemistry both involve the transfer of electrons. With the transfer of electrons comes oxidation and reduction. Be sure you know how to determine which element gets oxidized and which gets reduced in a redox equation by understanding how to determine oxidation numbers. Also, make sure you know how to balance redox reactions in both acid and base conditions. For electrochemistry, be familiar with voltaic cells in terms of their components, the flow of electrons, and the name for each half cell. Also know how to calculate E^{o}_{cell} and understand the difference between E^{o}_{cell} and E_{cell}. Finally, understand the Nernst Equation as well as how use it for calculating non-standard potentials.

CHAPTER 6

ELECTRONS

OVERVIEW

Electrons are a subatomic particle found outside the nucleus of the atom. The term 'outside the nucleus' is a very general expression that does not really tell us much about electrons. In this chapter we will discuss what is known about electrons in terms of the way in which they fill energy levels, and also the system that is set in place to pinpoint specific electrons.

1. LIGHT AND THE ELECTROMAGNETIC SPECTRUM

Before we start looking at electrons, we will first talk about light and the electromagnetic spectrum which usually goes hand and hand with the study of electrons. Some important terms that you will need to be familiar with are the following:

Speed of light [C] – Constant value for the speed of light in a vacuum equal to $3.0x10^8$ m/s

Wavelength [λ] – Light moves in a wave. The distance between crests is known as the wavelength

Frequency [v] – The number of waves that pass a fixed point per unit of time

These three terms are all related by the following formula: $C = \lambda v$

The electromagnetic spectrum shows the different forms of electromagnetic radiation in order of increasing wavelength and decreasing frequency, as illustrated below:

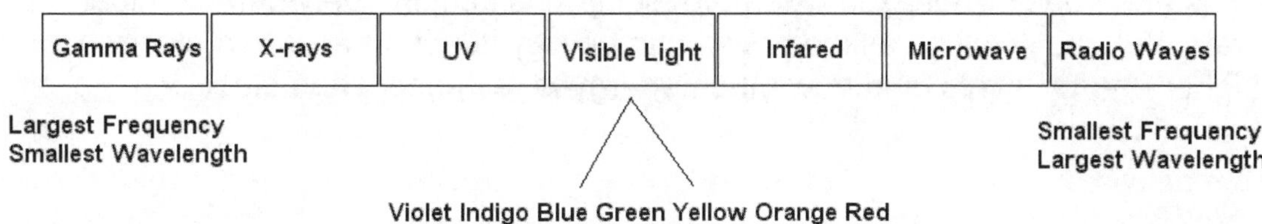

Be sure you have a general idea where each form of electromagnetic radiation is on the spectrum. Typical test questions involve asking which of the following choices have the highest frequency, lowest frequency, longest wavelength, etc. At a minimum you should know that radio waves have the smallest frequency/largest wavelength and that gamma rays have the largest frequency/smallest wavelength. Also, within the visible light region you should know where each color is. It is important that you are aware violet is the largest frequency/smallest wavelength and red is the smallest frequency/largest wavelength. A good way to remember the order is: ROYGBIV (Roy Gee Biv).

2. RYDBERG EQUATION

Before we knew as much as we do today about electrons, we used to think electrons revolved around the nucleus in spherical orbits. Even though we know that this is no longer true we still use this model as a simple means of explaining some topics. The Bohr Model of the atom, which looks like a dart board when drawn in 2 dimensions, is used to explain how electrons jump between energy levels. The rings in the Bohr model represent principal energy levels. When a beam of light hits an electron, the electron gets excited and moves to a higher principal energy level. At this point, the atom has absorbed the light energy. However, as this excited state is not a stable situation and the electron must come back down to a lower principal energy level. When the electron comes back down, energy is emitted. This is all illustrated in the following diagram.

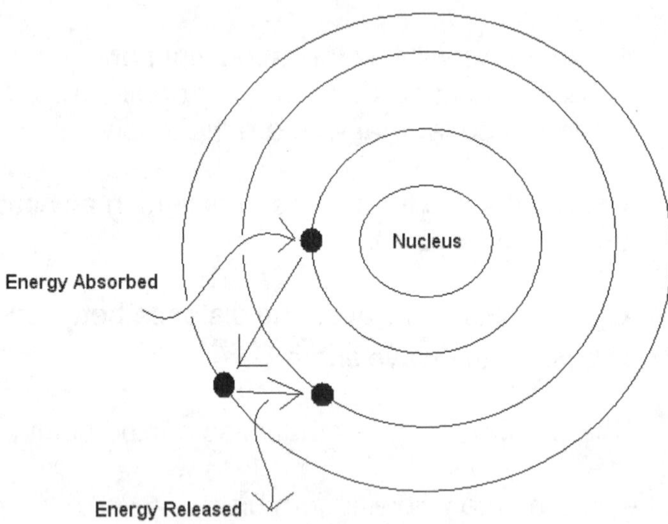

The types of questions you will see for this topic are questions involving the amount of energy that is absorbed or released when the electron is jumping between principal energy levels. You may well be asked how much energy is absorbed when electrons move to higher energy levels, and how much energy is released when electrons move to lower energy levels.

The way that we calculate the amount of energy is with the Rydberg equation. The Rydberg equation is:

$$E = -2.18\times10^{-18}\left(\frac{1}{n^2_{\text{lower}}} - \frac{1}{n^2_{\text{higher}}}\right)$$ (where n = principal energy level)

Example: How much energy is released when an electron moves from the 5th principal energy level back down to the 3rd principal energy level?

Step 1: Write the Rydberg Equation

$$E = -2.18\times10^{-18}\left(\frac{1}{n^2_{\text{lower}}} - \frac{1}{n^2_{\text{higher}}}\right)$$

Step 2: Plug in the principal energy levels from the question

$$E = -2.18\times10^{-18}\left(\frac{1}{3^2} - \frac{1}{5^2}\right)$$

Step 3: Solve

$E = -1.55 \times 10^{-19}$

**Interesting Fact: The reason why a sidewalk feels hot when the sun is beating on it is because the sunlight energy gets absorbed by the sidewalk atoms, causing electrons to move to a higher principal energy level. These atoms are not stable and the electrons are forced to move back to a lower energy level. In the process energy is released in the form of heat, which is why the sidewalk feels hot.

3. ELECTRON CONFIGURATION

In a separate chapter we talked about how to determine the number of electrons in a particular atom or ion. In this chapter, not only are we interested in the number of electrons, but also with where all of the electrons reside. Before we start talking about the way in which electrons fill, we should first discuss what they fill.

3a. What electrons fill

Principal energy level – A principal energy level is the broadest description of where an electron is. What is meant by principal energy level is the number in front of the electron configuration. The analogy that I always use for how to view the principal energy level is to think of a map of the United States. When trying to pinpoint where you live, you start with the overall state you live in to direct the observer's attention to the area in the United States that you live. To keep a running analogy as we study how electrons fill, let's consider the principal energy level to be the state of Delaware.

Subshell – Within a principal energy level, there are subshells. A subshell would be the equivalent of saying which county in Delaware you live in. Let's consider the subshell to be New Castle County. There are four different types of subshells you should know about; s, p, d and f. Knowing which principal energy level we are looking at determines how many subshells there are. In the first principal energy level there is only one subshell; named the 1s subshell. In the second principal energy level there are two subshells: 2 s and 2p. In the third principal energy level there are three subshells: 3s, 3p and 3d. In the fourth principal energy level there are four subshells: 4s, 4p, 4d and 4f.

You will not be asked about higher level subshells, but I hope you see that the pattern now is that the number of subshells is the same as the number of the principal energy level; so there will be 5 subshells in the 5th energy level etc.

Orbital – Within each subshell there are orbitals. An orbital would be the equivalent of saying which particular house in New Castle County is yours. An orbital defines the space or region around the nucleus where an electron actually resides. Each orbital can hold a total of two electrons each. As far as the shapes of these regions go; the shape of an s orbital is a sphere; a p orbital is like a dumbbell, and a d orbital is like a bimodal dumbbell. (You will not be responsible for knowing the shape of an f orbital.)

The total number of orbitals depends on which subshell we are talking about. Every s subshell has 1 orbital; every p subshell has 3 orbitals; every d subshell has 5 orbitals and every f subshell has 7 orbitals. Since there can be a total of two electrons in each orbital, this means that any s subshell can hold a total of 2 electrons, any p subshell can hold a total of 6, any d subshell can hold a total of 10, and any f subshell can hold a total of 14.

The following figure is a summary of the location of electrons in terms of principal energy level, subshells and orbitals:

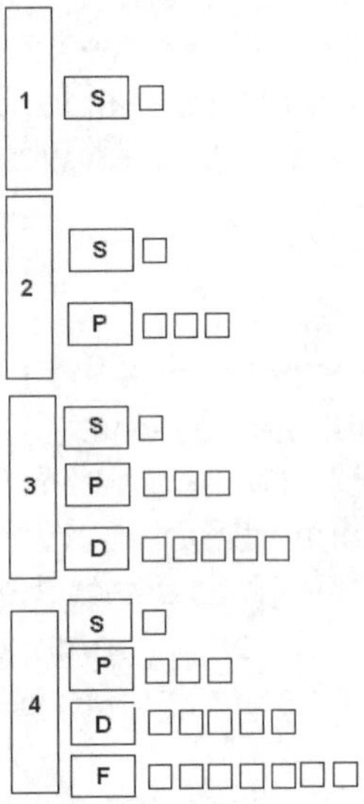

3b. How electrons fill

Now that we know how to describe the orbitals where electrons fill, we must discuss the order in which they fill. All electrons, regardless of which element they belong to, all fill in the same order, starting at 1 s and moving upwards. I now explain two different ways of remembering how electrons fill.

Approach #1:

The easiest way to remember the order of the subshells which electrons fill is using the following system:

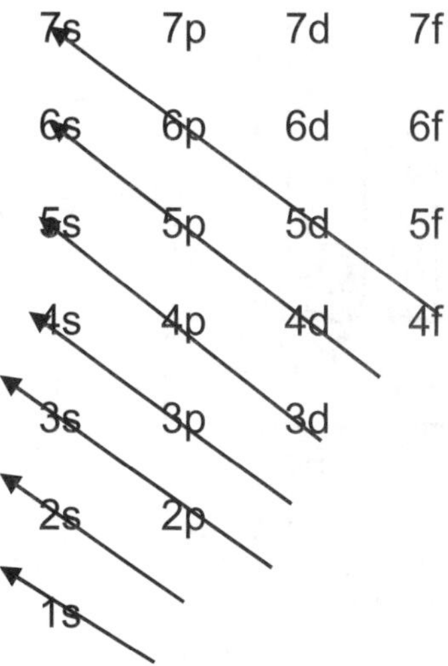

All you have to do is follow each arrow, starting at 1s, and when you get to the end of an arrow you move to the start of the next arrow. Using this figure to remember the order of how the electrons fill the subshells, as well as how many electrons can fit into each subshell (as discussed previously), you can successfully write an electron configuration.

Example: Write the electron configuration of Ar

Answer: $1s^2 2s^2 2p^6 3s^2 3p^6$

Explanation: Argon is atomic number 18, meaning it has 18 protons. There must be an equal number of electrons as there are protons, since it does not have a charge.

Start following the arrows and filling the subshells to the maximum amount of electrons possible until you get to the total 18. In the configuration as written, note that the numbers in front represent the principal energy level, the lower case letters represent the subshells, and the exponents represent how many electrons are in each subshell. By adding all the exponents, which are the numbers of electrons, you know you are finished when they sum to 18.

Approach #2:

You can determine the order in which electrons fill based solely on the periodic table. The periodic table can be broken into 'blocks' in which all of the elements in a particular block are filling the same subshell. Within each block, as you go from left to right, there is one additional electron in each successive element. The following periodic table is drawn to emphasize these different blocks:

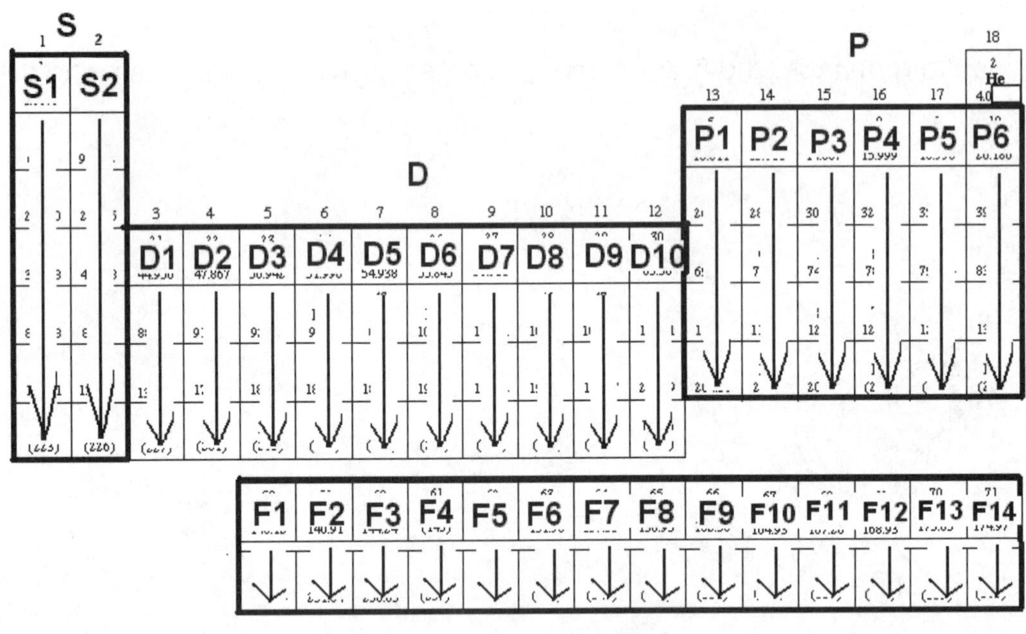

As you can see in the above figure, the periodic table actually reflects the electron configuration! As we know, any s subshell holds 2 electrons, and in the periodic table, the s block has two columns, one for each electron in the s subshell. Any p subshell can hold a total of 6 electrons, which is why there are 6 columns in the p block on the periodic table. The same goes for the d and f blocks, each of which have the same number of rows as electrons in that subshell.

The way that you should use this method is to first note where on the periodic table the element in question belongs. Then, start at the top of the periodic table and go horizontally to the right keeping track of every electron.

Example: Write the electron configuration for Fe

Answer: $1s^2 2s^2 2p^6 3s^2 3p^6 4s^2 3d^6$

Explanation: Starting at the top of the periodic table and going across a period to the right, you will notice the order in which the subshells are being filled. Notice how Fe is 6th element in the d block. This means it has 6 electrons in the d subshell, so its configuration must end with d^6. Then complete the configuration, we just write in the orbitals you passed over to get to Fe, filled with the max number of electrons as before.

Keep in mind that regardless of the approach you choose to use to obtain the electron configuration, you should get the same answer! For this reason, it is safe to use either approach. I would recommend picking the approach that you are most comfortable with.

3c. Shortcut

Writing electron configurations often get a bit tedious, especially for elements that have a large number of electrons. Worse still, you are more likely to make a mistake with long configurations.

In order to avoid this, chemists use a system to abbreviate the configuration. To use this, first identify the largest noble gas that comes <u>before</u> the element for which you are trying to write the configuration. For example, every single configuration for atoms with more than 18 electrons starts exactly the same way, namely $1s^2 2s^2 2p^6 3s^2 3p^6$ etc. Instead of writing all of this, you can simply write [Ar] instead, and then continue where that leaves off. By writing [Ar] we account for the first 18 electrons in the configuration, and can start at 19 instead of starting at 1. If the elements you are writing the configuration for is really big, you can use an even bigger noble gas as go from there.

Example: Write the electron configuration for As

Answer: [Ar] $4s^2 3d^{10} 4p^3$ (instead of having to write $1s^2 2s^2 2p^6 3s^2 3p^6 4s^2 3d^{10} 4p^3$)

3d. Exceptions

The electron configuration is the **ground state** of the atom. The ground state basically means that electrons will fill in the correct order, generally only moving to higher levels once lower levels are full. However it is important to note that there are some apparent exceptions to this rule. Some elements recruit electrons from certain subshells and add them higher subshells first, in order to make the higher subshell more stable. Usually when higher subshells are occupied before lower subshells are filled the atom would be in an **excited state**. However, since these elements are exceptions to the rules, even though they have electrons in higher subshells before lower ones are completed they are still considered to be in the ground state.

The two most popular elements that fall under this category are Copper and Chromium. Copper and Chromium are always the elements that instructors ask about, so be aware that when you see them you should automatically recognize that they are exceptions.

Copper: Electron Configuration $1s^2$ $2s^2$ $2p^6$ $3s^2$ $3p^6$ $\mathbf{4s^1}$ $\mathbf{3d^{10}}$

**The reason why the electron from the 4s orbital is taken by the 3d orbital is because the 3d orbital is only 1 electron away from becoming completely full, and d^{10} is more stable than d^9. The element becomes more stable by recruiting one electron from the 4s subshell first in order to complete a full 3d subshell.

Chromium: Electron Configuration $1s^2$ $2s^2$ $2p^6$ $3s^2$ $3p^6$ **$4s^1$ $3d^5$**

**The reason why the electron from the 4s orbital is taken by the 3d orbital is because the 3d orbital is one electron away from becoming exactly half full. This is also a stable arrangement, where d^5 is more stable than d^4. By recruiting electron from the 4s, the fifth d orbital in the 3d subshell can hold one electron making the atom more stable.

The other exceptions similar to the ones above are Niobium, Molybdenum, Ruthenium, Rhodium, Palladium, Silver, Platinum, and Gold. Be aware that these elements have exceptions. Typically you will not be tested on these exceptions, but it is important ot understand Copper and Chromium. And again, it is important to always check with your instructor if any of these other exceptions will be on a test!

4. QUANTUM NUMBERS

I like to think of quantum numbers as a means of pinpointing exactly where a specific electron is. As we have talked about previously, electrons are in certain orbitals within certain subshells, within certain principal energy levels. The best analogy to think of quantum numbers is the mailing system in the United States. If I wanted to send you a letter on your college campus, there must be a system set in place to ensure that the letter gets to you. First, starting off general, I would have to specify which complex on campus you live in. Then, I would have to say which building within the complex you live. I would then have to say which room within the building you live. Lastly, assuming that you have a roommate, I would have to identify your name specifically to make sure the letter gets to you. The same type of system exists for electrons and uses quantum numbers. Before we get into quantum numbers, I would first like to explain visually what is meant by electron configurations, in order to set the stage for quantum numbers.

4a. The orbital view of electrons

When we write electron configurations, we write a series of numbers and letters. The question is, what do these numbers and letters really refer to? They refer to the orbitals where electrons are to be found. As I stated earlier orbitals are depicted with boxes. The following figure is what electron configurations look like, using boxes for orbitals:

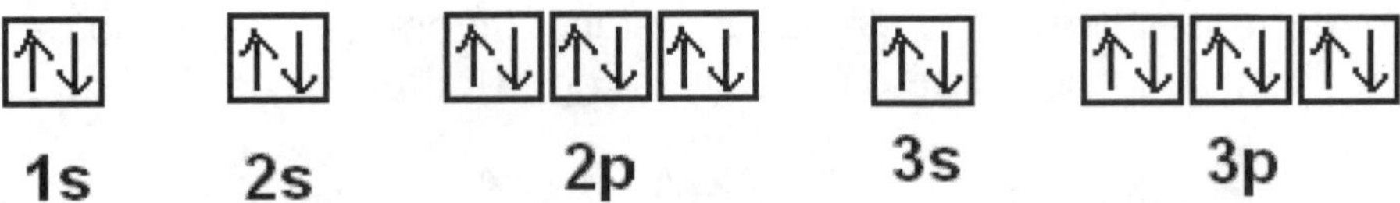

This would represent the electron configuration of $1s^2 2s^2 2p^6 3s^2 3p^6$. Each arrow indicates an individual electron. Each box is an orbital and can contain up to 2 electrons. There is 1 orbital in the s subshell, and 3 orbitals in the p subshell etc.

4b.What quantum numbers mean

There are 4 different quantum numbers; n, l, m_l, m_s. Let's explain each in turn:

n: Principal Energy Level

The principal energy level refers to the number in front of the subshell letter. It can be any integer starting at 1. The Principal energy level is the first quantum number and is used to designate which part of the electron configuration we should be looking at to find the one specific electron we are trying to find. As far as the analogy I introduced earlier, this would be the equivalent of the complex on campus which would be the first piece of information I would need to designate to send you a letter.

Example: n=2 refers to the 2nd principal energy level

l: Subshell

The subshell refers to s, p, d, or f. Instead of writing these letters, there is a system in place to designate which particular subshell we are talking about. The system is:

s =0
p =1
d =2
f = 3

Therefore, when specifying a subshell, you must use the number that corresponds to that subshell. This specifies which set of boxes within the principal energy level you are referring to. As far as the analogy goes, this would be the equivalent of stating which building within the complex you live.

Example: n=2, l =1 refers to the 2nd principal energy level, and the P subshell

m_l:Orbital

The orbital refers to which particular box you are talking about. To designate which box you are talking about, you use the interval – l to + l. This notation becomes particularly important for subshells that have more than one orbital. In the case of a p subshell for example, with l = 1, the orbitals would be designated from -1 to +1 as illustrated below.

-1	**0**	**+1**
↑↓	↑↓	↑↓

2p

Think of numbering the orbitals as if they were a number line. The central orbital is always zero and the numbers increase positively to the right, and negatively to the left as if it was a number line. This is the case for all subshells. As far as the analogy goes, the orbitals would be the equivalent of the room number. When you think about it, an orbital is basically the electron's room.

Example: n=2, l =1, m_l=-1 refers to the 2nd principal energy level, the P subshell, and the first orbital

m_s: Specific Electron

Each orbital holds 2 electrons max. In order to specify which specific electron out of the 2 we are talking about we use m_s. The up facing electron in the orbital is +1/2 and the down facing electron is -1/2. As far as our analogy, this would be the equivalent of using the name to specify which particular roommate in the room the letter is going to.

Example: $n=2$, $l=1$, $m_l=-1$, $m_s=+1/2$ refers to the 2nd principal energy level, the p subshell, the first orbital, and the up facing electron.

4c. The way that you will see this information asked on an exam

On an exam, you are most likely not going to be asked to point to the specific electron that a set of quantum numbers is being used to describe. Instead, you will see a lot of different multiple choice questions such as "which of the following quantum sets is incorrect?" and "what is the maximum number of electrons for the following sets of quantum numbers?" Let's go over how you would go about answering both types of questions.

Which of the following quantum sets is incorrect?

A quantum set is all of the quantum numbers in the following order $\{n, l, m_l, m_s\}$. The quantum sets that are not correct are the ones that describe an impossible circumstance. For instance, if there was a set of quantum numbers that refer to an f subshell in the 2nd principal energy level, this would be incorrect! As we know, there is no f subshell in the 2nd principal energy level. The typical things that you want to look out for when looking for incorrect quantum numbers are the following:

n – can be any number greater than 0
l – ranges from zero up to (n-1) only
m_l – must be in the range – l to + l (and can include $m_l = 0$)
m_s – Must be the value +1/2 or -1/2

If one or more of these rules is not met, you have incorrect set of quantum numbers.

Example: Which of the following is an <u>incorrect</u> set of quantum numbers?

[A] {2,1,-1,-1/2} [B] {4,3,2,+1/2} [C] {1,0,0,-1/2} [D] {3,3,-3,-1/2}

Answer: Choice D is the correct.

Explanation: Remember that l must be 1 less than n or lower. In choice D n =3 and l =3 which cannot happen. The reason why it can't happen is because the 3rd principal energy level has 3 subshells, and l is only allowed to be 0, 1 or 2. In other words, there is no f subshell in the 3rd principal energy level.
What is the maximum number of electrons for the following set of quantum numbers?

For these types of questions, just remind yourself what the set of quantum numbers is referring to. Once you make sense out of the quantum numbers you can answer any question in regards to the number of electrons it refers to. Keep in mind, for each additional quantum number that is provided, there are fewer and fewer possible electrons that it would be referring to. If you think about it logically, the more specific the quantum numbers get, the less electrons that can fit that set of quantum numbers. For instance, if only a principal energy level is given, there can be a lot of electrons that fit that general description. However, if all of the quantum numbers up to and including the m_s are given, then only 1 electron can have exactly that set of quantum numbers!

Example: What is the maximum number of electrons in {3,2}

[A] 2 [B] 3 [C] 10 [D] 14

Answer: Choice C is the correct.

Explanation: Remember that this set of quantum numbers refers to the 3rd principal energy level (n=3) and the d subshell (l = 2). The d subshell has 5 orbitals and so can have a total of 10 electrons.

SUMMARY

Electrons exist in principal energy levels outside of the nucleus. Be sure that you know how to calculate the amount of energy that is associated with a jump from one principal energy level to another using the Rydberg equation. Also make sure you are comfortable with electron configurations, the exceptions and the shortcut. Lastly, understand what quantum numbers describe as well as what would constitute an incorrect set of quantum numbers. Questions about material covering electrons are typically pretty easy and often serve as an excellent way of boosting test scores.

CHAPTER 7

EQUILIBRIUM CALCULATIONS

OVERVIEW

Chemical equilibrium is the state in which the forward chemical reaction proceeds at the same rate as the reverse chemical reaction. At equilibrium, there is no net change in the concentrations of reactants and products. What is interesting about chemical equilibrium is that it is dynamic, meaning it is constantly moving. It is incorrect to think that the reaction stops at equilibrium. The reaction does not stop, but rather the forward and reverse reactions are equal in rate which leads to no net changes in reactant or product concentrations. This unique characteristic about equilibrium leads to a variety of important topics that will be discussed in this chapter.

We start with LeChatelier's Principle which is to do with maintaining equilibrium when a change is implemented. After that we will talk about equilibrium constants and how to solve important problems relating to them. Also, the constant Q will be explained. Lastly, this chapter will conclude by explaining how to convert between two important equilibrium constants.

1. LE CHATELIER'S PRINCIPLE

Le-Chatelier's Principle is a method we use to explain how a system at equilibrium counteracts changes that are performed. Nature tries to work against any changes that have been done to the reaction at equilibrium. An excellent analogy that I like to use to explain LeChatelier's Principle is the concept of a seesaw. When there are two people on a seesaw who weigh the same as each other, the seesaw can remain perfectly horizontal. At this state the seesaw would be in equilibrium. If another person were to get on the right side of the seesaw, the once horizontal seesaw would go down on the right end and go up in the air on the left end. This addition of an extra person on the seesaw caused the seesaw to not be in equilibrium anymore. In order to counteract that change, and put the seesaw back in equilibrium, a person of the same new weight would have to get on the left side of the seesaw.

Relating the seesaw analogy back to chemistry, anytime there is a change implemented on a system at equilibrium, nature must do something to offset this change. The types of changes that can occur to a system in equilibrium include increasing or decreasing the concentration of a reactant or product; changing the volume; changing the pressure, or changing the temperature. All of these changes disrupt the equilibrium and cause it to shift. Depending on the change this shift will determine the way in which the system will react in order to offset the change. The following sections address the way in which the reaction responds to each type of change.

1a. Increasing or decreasing reactant or product concentrations

In general, whenever something is added to one side of a reaction, it causes a shift to the opposite side of the reaction to counteract that shift. Also, if something is taken away from one side of the reaction, the reaction shifts towards the side that lost something, in order to replenish what was lost.

Example: C_3H_8 (g) + 5 O_2 (g) ↔ 3 CO_2 (g) + 4 H_2O (g)

Increase [O_2]: Causes shift to the right, so more products are formed
Decrease [H_2O]: Causes shift to the right, so reactants react to replenish the H_2O

1b. Increasing or decreasing pressure

When changing pressure, the most important thing to look for is the total moles of gas on each side of the reaction. When the pressure increases, the reaction will shift to the side with the fewer total moles of gas. When decreasing the pressure, the reaction will shift to the side with the greater total moles of gas. An easy way of looking at this is to think of it as pushing everyone in a crowded room together will make them want to move to the other side of the room to counteract the increase in pressure.

Example: C_3H_8 (g) + 5 O_2 (g) ↔ 3 CO_2 (g) + 4 H_2O (g)
(6 moles of gas) (7 moles of gas)

Increase Pressure: Causes shift to the left, so more C_3H_8 and O_2 are formed
Decreasing Pressure: Causes shift to the right, so more CO_2 and H_2O formed

1c. Increasing or decreasing volume

Volume also has to do with the number of moles of gas on each side of the reaction. When you think about it, the rules for volume changes are actually the exact opposite of those for pressure changes. The reason for this is because when you increase the pressure you decrease the volume. Therefore, an increase in volume (really a decrease in pressure), shifts to the side with the greater total moles of gas. A decrease in volume (or an increase in pressure), shifts to the side with the fewer total moles of gas.

**Note that the best way to think of volume is to just think of it in terms of pressure. This way you only have to memorize one set of rules out of the two and you will eliminate the chance of confusing them.

Example: C_3H_8 (g) + 5 O_2 (g) ↔ 3 CO_2 (g) + 4 H_2O (g)
(6 moles of gas) (7 moles of gas)

Increase Volume: Causes shift to the right, so more CO_2 and H_2O formed
Decrease Volume: Causes shift to the left, so more C_3H_8 and O_2 are formed

1d. Increasing or decreasing temperature

For temperature, it depends on whether the reaction is exothermic or endothermic. Remember an exothermic reaction gives off heat as a product and has a $-\Delta H$. An endothermic reaction takes in heat as a reactant and has a $+\Delta H$. The key to temperature questions is to write the word heat into the reaction as a reactant if it is endothermic, or as a product if it is exothermic. This way you can think of heat as an actual component. When increasing the temperature, it shifts from the side with the heat. When decreasing the temperature it shifts towards the side with the heat.

Example: $C_3H_8\,(g) + 5\,O_2\,(g) \leftrightarrow 3\,CO_2\,(g) + 4\,H_2O\,(g)$ $\Delta H = -200$ kJ

First, write this as: $C_3H_8\,(g) + 5\,O_2\,(g) \leftrightarrow 3\,CO_2\,(g) + 4\,H_2O\,(g)$ + Heat

Increase Temperature: Causes shift to the left, so product heat is consumed
Decreasing Temperature: Causes shift to the right, so more heat is generated

2. WRITING EQUILIBRIUM CONSTANTS

There are a few important things that you must be aware of when writing equilibrium constants. The first thing is that equilibrium constant expressions are always written with the concentrations of products in the numerator and the concentrations of reactants in the denominator. The coefficient in front of a compound is written as the exponent on that compound's concentration in the equilibrium constant expression. Concentrations of solids and liquids are considered constant and are not written in the equilibrium constant expression. So the equilibrium expression includes only aqueous phases and gas phases, and can be written in terms of concentration (K_c) or pressure (K_p).

For example, in the reaction: $2\,A(aq) + 3\,B(aq) \leftrightarrow C(aq) + 2\,D(aq)$

We write the generic Equilibrium constant as:

$$K_C = \frac{[C][D]^2}{[A]^2[B]^3}$$

**Important notes– If there is a solid or liquid in the equation be sure to exclude it when calculating the equilibrium constant. Also, K_c is not written with units: it is just a number.

3. SOLVING EQUILIBRIUM EQUATIONS

Questions concerning equilibrium equations usually give us a value for Kc and the concentrations of some of the components, but we have to find the concentration of one of the unknown reactants or products. My best advice for solving equilibrium equations is to first write the equilibrium constant expression. Write x to represent the concentration of the unknown component. Then determine all the given that will be used to represent the remaining products and reactants. Finally plug those values into the equilibrium constant expression and then solve for x.

Solving equilibrium equations is similar to solving weak acid/base problems. However, there is one important thing to note. The coefficients in front of the compounds in the chemical reaction become a little tricky in equilibrium questions because they appear to be taken into account twice. What I mean by this is that not only is the coefficient used as the exponent in the equilibrium constant, but it is also used as the coefficient for the variable that is taking the place of the compound.

For example, you will notice in the following reaction; $Cl_2(g) \rightarrow 2\ Cl(g)$, that the product Cl has a coefficient of a 2 in front of it. If the concentration of Cl is the unknown value, when writing the equilibrium constant expression it gets written as $(2x)^2$. Although it appears that the coefficient gets taken into account twice, it really does not. What happens here is that 2x is used to denote the concentration of Cl. So when this value is plugged back into the equilibrium equation, 2x itself gets raised to the power of two.

Example: If 0.5 moles of CCl_4 gas is sealed in a 3 L beaker, the following reaction occurs: $CCl_4(g) \leftrightarrow C(s) + 2\ Cl_2(g)$. What will the concentration of CCl_4 and Cl_2 be when the reaction has reached an equilibrium where Kc = 0.0001?

Step 1: Write the reaction:

$CCl_4(g) \leftrightarrow C(s) + 2\ Cl_2(g)$.

Step 2: Write the equilibrium constant expression:

$$K_C = \frac{[Cl_2]^2}{[CCl_4]}$$

Step 3: Determine initial concentration of CCl_4 using the information given:

$$M = \frac{\text{moles}}{L}$$
$$= \frac{0.5\text{ moles}}{3\text{ L}}$$
$$= 0.167\text{ M}$$

Step 4: Use this value to keep track of the concentrations of the other components before the reaction starts and at equilibrium. To do this, use X to denote how much of the reactant gets converted to product at equilibrium. Also, ignore C as it is a solid:

CCl_4 $\leftrightarrow$	C +	$2\ Cl_2$
0.167		0
0.167-x		2x

Step 5: Plug values into equilibrium constant expression:

$$0.0001 = \frac{[2x]^2}{[0.167 - x]}$$

Step 6*: Assume that x is very small compared with 0.167, so it can be ignored. Then, condense and solve for x:

$$0.\ 0001 = \frac{4x^2}{[0.167]}$$

$$4x^2 = 1.67 \times 10^{-5}$$

$$x = 2.04 \times 10^{-3}$$

Step 7: Plug in x value to the concentrations illustrated in the last row of the table in step 4 to calculate the concentrations of the other compound(s):

$[CCl_4]$ = 0.167 – 0.00204 = 0.165 M

$[Cl_2]$ = 2x = 2(0.00204) = 4.08×10^{-3} M

*Step 6 is very important. These calculations are a lot easier if we can ignore any appearance of x in the denominator, as we did above. However, if x turns out to be big we cannot do this, and instead will need to manipulate the expression in step 5 into a full quadratic equation. An easy way to check is to see if $[CCl_4] - x \approx [CCl_4]$. If it is, we can justify this short-cut. Note that in this example, $[CCl_4] - x = 0.165 \approx 0.167$, so ignoring x was justified.

4. Q

A lot of students who take introductory chemistry get confused when they see Q, or the quotient of concentrations. Q tells us exactly where a reaction is in terms or whether it is not yet at equilibrium, at equilibrium, or if it passed equilibrium. In fact, it is best to think of Q as a snapshot in time. Q is not a true equilibrium constant, but is calculated in exactly the same way. In other words, we write the equilibrium constant expression and plug in the values for the concentrations.

If K is smaller than Q, the reaction is actually beyond equilibrium. This causes a shift to the left causing the concentration of reactants to decrease and the concentration of the products to increase.

If K is larger than Q, the reaction is not yet at equilibrium. This would cause a shift to the right causing the concentration of reactants to decrease and the concentration of the products to increase.

If K is equal to Q, then the reaction is in fact at equilibrium

5. K_P AND K_C

Kp and Kc are two important equilibrium constants and it is important to be able to convert one to the other. Kp is the equilibrium constant calculated in terms of pressures of gas components and Kc is the equilibrium constant in terms of concentrations. In order to convert between the two constants use the following formulas:

$$K_C = K_P(RT)^{-\Delta n_{gas}}$$

$$K_P = K_C(RT)^{+\Delta n_{gas}}$$

***Easy way of remembering this: The only difference between the two formulas is the positive and negative exponent. A good way of remembering that the K_p formula has the positive exponent is by thinking “p stands for positive”

The only thing that gets a little tricky with these equations is the exponent Δn. Δn stands for change in the moles of gas on both sides of the reaction. The way to calculate Δn is to use the formula:

Δn = Total moles of product gases – Total moles of reactant gases

This is how to calculate Δn, but do not forget to then obey by the + or – in front of the Δn exponent in the formulas.

Example: If the K_p value is 6.2 x 10^{-6} for the following reaction at T = 200 K, what is the K_c value? (Use R = 0.0821 $L.atm.mol^{-1}.K^{-1}$)

$N_2(g) + 3 H_2(g) \leftrightarrow 2 NH_3(g)$

Step 1: Recognize that we must use the formula that converts between K_p and K_c since we are given K_p and must find K_c.

Step 2: Write the formula for converting K_p to K_c

$$K_C = K_P(RT)^{-\Delta n_{gas}}$$

Step 3: Calculate Δn

Δn = Total moles of product gases – Total moles of reactant gases
$\Delta n = 2 - 4$
$\Delta n = -2$

Step 4: Plug information into formula and solve

$$K_C = 6.2\times10^{-6}[(0.0821)(200)]^{-(-2)}$$

$$K_C = 1.67\times10^{-3}$$

SUMMARY

In this chapter we discussed chemical equilibrium and the calculations that you will need to know for the topic. LeChatelier's Principle is an important concept that is used to explain how a system counteracts changes in order to re-establish equilibrium. Be sure you know each type of change and the direction of the shift it causes. Make sure you know how to write an equilibrium constant expression and solve it. Don't forget about how the coefficients seem to appear twice, and don't let it confuse you. Understand what Q tells you and view it as a snapshot that tells you where the reaction is compared to the equilibrium state. Lastly, make sure you know the equation to convert between K_p and K_c. If you can do these operations, you should have no problem with chemical equilibrium questions.

CHAPTER 8

GAS LAWS

OVERVIEW

Most of the chemistry that we do under gas laws involves ideal gases. What is meant by the term 'ideal' gas is that the gas behaves how it is expected to behave under standard conditions. However, this is not an ideal world. The gases that chemists work with in real life do not behave ideally and this makes our lives a lot more complicated. Luckily, in introductory chemistry, you will mostly be working with ideal gases.

In this chapter we will talk about the important equations that you need to know for studying gases as well as simple techniques for using them. This chapter will identify the equations, explain when to use the equations, and offer examples of how to do this. By the end of this chapter you will be able to recognize gas law problems, know which equation to use, and know exactly how to use that equation for the problem at hand.

1. IDEAL GAS LAW

Equation: $PV = nRT$

Components:
- P = Pressure (in atm)
- V = Volume (in L)
- n = number of moles
- $R = 0.0821 \frac{L \cdot atm}{mol \cdot K}$
- T = Temperature (in K)

The ideal gas law is an extremely easy and straightforward equation to use. The most important thing to remember about the ideal gas law is that you must use the proper units for each component. Since the gas constant R is usually given in units of $\frac{L \cdot atm}{mol \cdot K}$, it is imperative that you use these units for Temperature, Pressure and Volume so they will cancel out. For example, since there is L in the units of the gas constant, the volume that you plug into the equation must be in L and not mL. Also, instructors love to test you here by giving you units that need to be converted! The way that you should do this is using dimensional analysis to make sure all of the starting units cancel and that you are left with the unit that you are looking for.

The typical question that you will be asked using this equation is to solve for one of the variables after being given the rest of the variables. Keep in mind that there are 4 different variables (R is a constant and not a variable), meaning there are 4 different questions that can be asked using this equation. Regardless of this, treat every question the same way. First you need to recognize that the question will be requiring you to use the ideal gas law. You can do this when you see that you have a question that simply states 3 gas related variables and asks you to solve for the fourth. Simply circle the variable that they are asking for and then reorganize the equation algebraically to solve for that variable. Lastly, plug in the values that you are given while being sure that they are all in the correct units, and solve.

Example: An ideal gas takes up .5mL at a temperature of 25°C. The pressure of the tank is 780 torr. How many moles of gas are present?

Step 1: Recognize that this question involves the use of the ideal gas law because you are given a volume, temperature, and pressure, while being asked to solve for moles.

Step 2: Write the ideal gas law and circle the variable you are being asked to solve for:

$$PV = \boxed{n}\,RT$$

Step 3: Rearrange the equation to solve for n

$$n = \frac{PV}{RT}$$

Step 4: Plug in values:

$$n = \frac{(780\ \text{torr})(0.5\ \text{mL})(\text{mol} \bullet \text{K})}{(0.0821\ \text{L} \bullet \text{atm})(25\ ^{\text{o}}\text{C})}$$

**note that the units for R is $\frac{L \bullet atm}{mol \bullet K}$ and that is why the units are written as above

Step 5: Use dimensional analysis to convert to all of the proper units:

$$n = \frac{(780\ \text{torr})(1\ \text{atm})(0.5\text{mL})(1\ \text{L})(\text{mol} \bullet \text{K})}{(760\ \text{torr})(1000\ \text{mL})(0.0821\ \text{L} \bullet \text{atm})(298\ \text{K})}$$

Step 6: Be sure all units cancel and that you are left with moles and solve:

$$n = \frac{(780\ \text{torr})(1\ \text{atm})(0.5\ \text{mL})(1\ \text{L})(\text{mol} \bullet \text{K})}{(760\ \text{torr})(1000\ \text{mL})(0.0821\ \text{L} \bullet \text{atm})(298\ \text{K})} = 2.10 \times 10^{-5}\ \text{moles}$$

2. COMBINED GAS LAW

Equation: $\frac{P_1V_1}{T_1} = \frac{P_2V_2}{T_2}$

Components: P = Pressure
V= Volume
T = Temperature
"1" and "2" indicate values before or after the change, respectively

The combined gas law is used when there is a physical change in the of the gas. Typical phrases that you should recognize as clues for using the combined gas law are ones such as, "when the temperature is dropped to…", "the volume is compressed to…", or "the pressure is increased to…" These types of phrases should act as a flag that help you recognize that the combined gas law is needed to solve the problem.

The one important thing to keep in mind for the combined gas law is that the units need to be the same for both conditions. In other words, if you are starting with a volume in mL then you must end with a volume in mL. Instructors might try to trick you by giving you a volume in mL but asking for a final volume in L. If this happens, be sure to realize that the answer you obtain is in mL since the starting volume is in mL; and that you need to convert mL to L for your final answer.

Remember that there are two forms of each variable, one for the initial condition and one for the final condition, so there are six variables total. The best way to go about solving combined gas law problems is to keep track of all six variables, and the best way to do this is to write down everything that is given in the word problem in a small table. Identify all of the variables you are given, and which one you need to obtain. Then rearrange the combined gas law to solve for that variable. You can do this by cross multiplying the two fractions and using algebra to calculate the unknown variable.

Example: One mole of gas at 300 K takes up 3 L when the pressure is 2.0 atm. If the temperature is dropped to 250 K what volume must the gas be compressed to in order to achieve a pressure of 6.0 atm?

Step 1: Recognize that this problem involves the combined gas law because of the phrase "What volume must the gas be compressed to"

Step 2: Write the combined gas law equation

$$\frac{P_1V_1}{T_1} = \frac{P_2V_2}{T_2}$$

Step 3: Write out all information given:

$P_1 = 2.0$ atm $\quad P_2 = 6.0$ atm
$V_1 = 3$ L $\quad V_2 = ?$
$T_1 = 300$ K $\quad T_2 = 250$ K

Step 4: Rearrange combined gas law to solve for V_2:

$$V_2 = \frac{P_1V_1T_2}{T_1P_2}$$

Step 5: Plug in information:

$$V_2 = \frac{(2.0\,\text{atm})(3\,\text{L})(250\,\text{K})}{(300\,\text{K})(6.0\,\text{atm})}$$

Step 6: Solve:

$V_2 = 0.833$ L

3. CALCULATING DENSITY AND MOLECULAR WEIGHT

Equation for calculating density: $d = \frac{(P)(MW)}{RT}$

Equation for calculating molecular weight: $MW = \frac{dRT}{P}$ or $\frac{wRT}{PV}$

Components:
P = Pressure
MW = Molecular weight
T = Temperature
d = Density
w = Weight
V = Volume
R = gas constant ($0.0821\ \frac{L \bullet atm}{mol \bullet K}$)

Calculating density and molecular weight are usually pretty straight forward. You will be able to recognize these types of questions because they will flat out ask you for either the molecular weight or the density. The hardest part is probably just memorizing the equations. For the density equation I think of P(MW) as the car BMW. For molecular weight, I remember dRT/P as “dirt over pee.” You will notice that the second equation for molecular weight is the same exact thing as the first only that instead of writing density, the term is just expanded as w/v - which is density.

Example: Calculate the molecular weight of the gas Kr at a pressure of 1 atm, density of 7 g/L and a temperature of 300 K

Step 1: Recognize that we need a molecular weight so we write the MW equation:

$$MW = \frac{dRT}{P}$$

Step 2: Plug in information given:

$$MW = \frac{(7\,g)(0.0821\,L \bullet atm)(300\,K)}{(L)(mol \bullet K)(1\,atm)}$$

Step 3: Solve:

MW = 172 g/mol

4. GRAHAM'S LAW OF EFFUSION

Equation: $\frac{t_2 l_1}{t_1 l_2} = \sqrt{\frac{MW_2}{MW_1}}$

Components:
- t = time
- l = length
- MW = Molecular weight
- "1" and "2" indicate values for the 1st or 2nd gas, respectively

Graham's law of effusion is used to compare the rate of escape of two different gases. In general, lighter gases effuse quicker than heavier gases; meaning the larger the molecular weight the slower the gas. You will notice both time and length appear in this equation. Typically, a question will only involve one of these two variables, and will ask for either a difference in time or a difference in distance while the other is held constant.

It is important to notice that with time as a variable, the same gas is written in the denominator of both sides of the equation and the other gas is written in the numerator of both sides of the equation. In other words, each gas remains on the same part of the fraction for both sides of the equation when using time. For length, it is the opposite. The first gas is written in the numerator on the first side of the equation and the denominator of the second side of the equation. Be aware of this difference and keep in mind which variable the question is providing you with.

When answering questions about the rate of effusion between two gases you will always be using Graham's law of effusion. I highly recommend keeping track of each gas by assigning the first gas as gas 1 and the second gas as gas 2. This will be important when you need to plug the numbers into the appropriate part of the fraction.

Example: If it takes 1.2 hours for 5 % of Xe gas to effuse through a hole, how long will it take for the same amount of Ne to effuse from the same hole?

Step 1: Recognize that it is a Graham's law of effusion question because it is asking about time of effusion of two different gases

Step 2: Write the effusion equation, ignoring l values (since these are kept constant)

$$\frac{t_2}{t_1} = \sqrt{\frac{MW_2}{MW_1}}$$

Step 3: Assign each gas a number

Ne = gas 1
Xe = gas 2

Step 4: Plug in information (take the MW of the gases from the periodic table)

$$\frac{1.2}{t_1} = \sqrt{\frac{131.3}{20.18}}$$

Step 5: Solve

Time = 0.47 hours (or 28 minutes)

**Note that you can convert to minutes by multiplying hours by 60

5. DALTON'S LAW

Equation: $P_{total} = P_{gas1} + P_{gas2} + P_{gas3}\ldots$

Components: P_{total} = Total pressure
P_{gas} = Partial pressure of each gas

Dalton's law states that the total pressure of any container is equal to the sum of all the partial pressures of the gases present in the container. The way that you will most likely see Dalton's law is in a question in which two separate containers of gas are combined. First you would want to use the combined gas law since a physical change occurs as the two containers are joined. This will leave you with a partial pressure of each gas involved. Once you obtain the partial pressure of each gas, you simply add them together to get the total pressure of the resulting container according to Dalton's law. The only thing that you must watch out for is that all of the gases have the same units.

Example: A 3 L container with 400 torr of Neon is separated from a 5 L container with 600 torr of Xenon. Once the two containers are joined by removing the blocker, what will be the resulting total pressure of the new joined container?

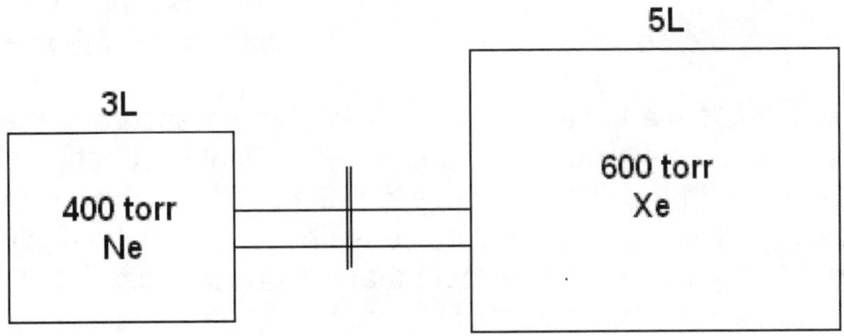

Step 1: Recognize that we will be using Dalton's law in this problem since it is asking for a total pressure and there is more than one gas involved

Step 2: Determine the partial pressure of Ne once the two containers are connected using the combined gas law (but ignoring temperature, since this does not change):

$P_1V_1 = P_2V_2$
$(400 \text{ torr})(3 \text{ L}) = (P_2)(3 + 5 \text{ L})$**
$P_2 = 150 \text{ torr}$

**Note V_2 is 8 L because a 3 L container is combined with a 5 L container to make 8 L

Step 3: Determine the partial pressure of Xe once the two containers are connected using the combined gas law (again ignoring temperature):

$P_1V_1 = P_2V_2$
$(600 \text{ torr})(5 \text{ L}) = (P_2)(8 \text{ L})$
$P_2 = 375 \text{ torr}$

Step 4: Use Dalton's Law to find the total pressure:

$P_{total} = P_{gas1} + P_{gas2}$
$P_{total} = 150 \text{ torr} + 375 \text{ torr}$
$P_{total} = 525 \text{ torr}$

6. MOLE FRACTION

Equation: $P_a = (X_a)(P_T)$

Components: P_a = Partial pressure of gas a
X_a = Mole fraction of gas a
P_T = Total pressure

The Mole Fraction X_a by the equation above is $\frac{P_a}{P_T}$. The easiest way to think of mole fraction is part over total. It is a fraction that consists of the pressure of the particular gas that we are talking about over the total pressure. Typically, questions involving mole fraction will provide you with the total pressure and the partial pressure and ask for the mole fraction of the gas. The equation makes sense because it is basically taking the total pressure and multiplying it by the percent that the gas is present out of that total and that value leaves us with the amount of pressure that gas is responsible for. The only thing you have to watch out for is that the units for both pressures are the same.

Example: What is the mole fraction of oxygen in the air if the partial pressure of oxygen is 25 torr on a day that the atmospheric pressure is 1atm?

Step 1: Recognize this requires the mole fraction, so write the mole fraction equation:

$P_{O2} = (X_{O2})(P_T)$

Step 2: Make sure units for the two pressures are the same

$$25\ \text{torr}\left(\frac{1\ \text{atm}}{760\ \text{torr}}\right) = 0.033\ \text{atm}$$

Step 3: Plug in information and solve:

0.033 atm = (X)(1atm)
X = 0.033

SUMMARY

Gas laws are relatively straight forward problems. They all involve plugging values into an equation. This chapter laid out the equations that you will be using for ideal gas problems as well as explained how to recognize which equations to use, and when.

Recognizing which equation to use is only half the battle. You also have to make sure that you know how to use the equations and what important things to look out for in each case. In equations that do not involve the gas constant you are free to use any unit that is given as long as you are consistent and use that same unit throughout the whole equation and in the answer.

CHAPTER 9

KINETICS

OVERVIEW

Kinetics has to do with the rate of chemical reactions. Kinetics can at times seem very tricky because it is often explained using calculus. However, in introductory chemistry, you will not be responsible for calculus. Even if your instructor presents the material using calculus, it is more so that you can see how calculus can be used but you will not have to use it. In this chapter we will discuss the important aspects of kinetics as well as look into techniques in which calculus can be avoided. We will start our discussion with the kinetic rate equation.

1. KINETIC RATE EQUATION

The kinetic rate equation is what we use to solve for the rate of decomposition of the reactants or the rate of appearance of the products. The kinetic rate usually accounts for most of the questions that are asked on tests covering kinetics. The kinetic rate equation is defined as:

$\text{Rate} = k[A]^a[B]^b$

The capital letters A and B in the equation stand for the concentration of the reactants. The lower case letters a and b stand for the order of each reactant. Sadly however, these are usually NOT the coefficients of each reactant in the equation! Instead we have to calculate them, as explained below. The orders of each reactant a or b tell us how each reactant contributes to the overall rate of the reaction based on the principle that reaction rate is proportional to reactant concentration. The last variable in the equation, k, is the kinetic rate constant.

In order to solve the rate equation for a given specific reaction we must calculate each variable by performing the following steps:

1) First solve for the order of each reactant by using the initial concentrations and initial rates found in the table provided.

2) Once the orders are calculated, plug those values into the general rate equation

3) Next we must solve for k using one of the trials in the table provided. By plugging in the values that correspond to one specific trial, we can solve for k.

4) Once we have solved for the orders, and for k we have our kinetic rate equation

Now let's look a little closer at each step in the process:

1a. Determining the order of each reactant

To determine the order of each reactant we must use the values that will be provided for us in the table given. For example:

Trial	[A]	[B]	Initial rate
1	2	8	2.0×10^{5}
2	2	4	1.0×10^{5}
3	9	5	9.0×10^{-3}
4	3	5	1.0×10^{-3}

We will start by calculating the order of reactant A. In order to do this we must pick two trials in which [B] (meaning the concentration of reactant B) is constant. By doing this, we will not have to worry about solving an equation with 2 unknowns. As you can see, in trials 3 and 4, [B] is the same.

Next calculate the ratio of concentrations of [A] in trials 3 and 4 and the ratio of the initial rates of trials 3 and 4.

$$\frac{[A]\text{in trial}3}{[A]\text{in trial}4} = \frac{9}{3} = 3 \qquad \frac{\text{Initial rate in Trial}3}{\text{Initial rate in Trial}4} = \frac{9.0 \times 10^{-3}}{1.0 \times 10^{-3}} = 9$$

Now, the order of reactant A will be the power to which the concentration ratio must be raised to get the initial rate ratio. So once we determine the ratios as above, we ask ourselves "to what power do we have to raise the concentration ratio, in order to make it equal the ratio of initial rates?" This value will be the order for the reactant.

In other words, we need to figure out which value of n will make $3^n = 9$
Answer: n = 2.

So the order for reactant A is 2.

To solve for the order of reactant B we do the same thing. First find the two trials in which reactant A is constant. From there we will again compare the concentration ratio to the ratio of initial rates and ask ourselves what power will make the ratios equal to each other. This value is the order for reactant B.

$$\frac{[B]\text{in trial}1}{[B]\text{in trial}2} = \frac{8}{4} = 2 \qquad \frac{\text{Initial rate in Trial}1}{\text{Initial rate in Trial}2} = \frac{2.0 \times 10^{-5}}{1.0 \times 10^{-5}} = 2$$

So now solve $2^n = 2$:
n = 1

So the order for reactant B is 1.

1b. Determining the rate constant k

At this point we now have the order of both reactant A and B. We can then plug these into the rate equation and be one step closer to solving our rate equation. In other words we now have Rate = $k[A]^2[B]^1$ but we still need to find the rate constant.

The way that we solve for the rate constant is to select all of the values from one trial from the table and plug these into the half-finished rate equation above. Technically, you can plug any trial into the rate equation to obtain k, but I personally like to always plug in the first trial just to make things easier on ourselves.

So, look at the first trial and plug in the concentrations and the initial rate into the equation that we have so far from trial 1.

$$2.0 \times 10^5 = k[2]^2[8]^1$$

We now have figured out every variable except k, so we can finally figure out k:

$$k = \frac{2.0 \times 10^5}{[2]^2[8]^1} = 6250$$

We can now write our final kinetic rate equation by plugging in all the values calculated:

$$\text{Rate} = 6250[A]^2[B]^1$$

**Note: we still need to find the units of k, but we'll talk about this below.

1c. The overall order of the reaction and units

We saw above how to determine the order of each reactant in the rate equation. In order to determine the overall reaction order we simply add the individual orders of the two reactants. The example above is a third order reaction. This is because reactant A is a second order reactant and reactant B is a first order reactant, so adding the two together makes it a third order reaction overall. Once you determine the overall order of the reaction, you will be able to determine the units for k. The following is a list of the units that correspond to each order reaction:

1st order: $\frac{1}{s}$ or s^{-1}

2nd order: $\frac{1}{M.s}$ or $M^{-1}s^{-1}$

3rd order: $\frac{1}{M^2s}$ or $M^{-2}\ s^{-1}$

So, the complete rate equation here is: Rate = $6250\ M^{-2}.s^{-1}[A]^2[B]$

2. REACTION MECHANISMS

The big picture in the case of kinetics is determining something called a reaction mechanism. In the industries, for example, certain reactions are used to fuel other reactions. In other words a product that is produced in one reaction is used as a reactant in another. More generally, most chemical reactions occur as a series of individual chemical steps. The overall system of reactions or steps is known as a reaction mechanism. In all reaction mechanisms, each step occurs at a different rate. The slowest step is the reaction that we concentrate on when talking about the overall rate. The reason for this is that a reaction can only go as fast as its slowest step.

**Analogy: Let's say you are running in a relay race. No matter how fast you might zip around the track and smoke every other opponent, your other team members might not be as fast as you are. In a team, you are each depending on one another, and your team is only as good as your weakest or slowest person. The way to assess the ability of your team is to focus on the slowest person.

Therefore relating this back to chemistry, a reaction can only proceed as fast as its slowest reaction. The slowest reaction is the **rate determining step.** In kinetics we are interested in knowing how fast a mechanism can proceed. So the way that we do this is by writing a kinetic rate equation using the reactants of the slowest step.

The advantage of using the rate determining step to write the kinetic rate equation is that we do not need to figure out the orders as we did above. Instead, we simply use the coefficients of each reactant shown in the rate determining step as the orders we want!

Example: What is the rate law (kinetic rate equation) for the following reaction: 2A + B + D → E, if it occurs in the steps shown below?

(slow) 2 A + B → C
(fast) C + D → E
(net) 2 A + B + D → E

Answer: Rate = $k[A]^2[B]$

**Note: the shortcut of using the coefficients directly to build the rate equation ONLY works if we are explicitly given the reaction rate determining step. If not, we have to calculate the orders as above.

2a. Important terms

There are a couple important terms when it comes to reaction mechanisms that are important for you to be familiar with. The following terms are illustrated using the following example reaction mechanism:

(slow) 2A + B → C + D
(fast) C + E → B + D

(net) 2A + E → 2D

Reaction intermediate - A reaction intermediate is a compound that is produced in one reaction step as a product then used in a late reaction step as a reactant. In other words, it is something that is first produced by the mechanism and then used in the mechanism. Reaction intermediates are used up in the mechanism and do not appear in the net equation. A reaction intermediate in the above example is C.

Catalyst – A catalyst is a compound that is present in a reaction mechanism and speeds up the mechanism. The catalyst starts as a reactant and then becomes a product, but it does not appear in the net reaction. So the catalyst above is B.

3. INTEGATED RATE LAWS

Everything that we have done so far involves the rate at which chemical reactions proceed. The calculus that we have so far avoided involves taking derivatives from the plot of concentration versus time. A derivative gives the rate of any point on a plot by drawing a tangent to the plot at that point and then taking the slope of that tangent to measure the rate. Integrated rate laws, by contrast, measure the reaction rates by a more accurate approach. Instead of measuring the slope of a line tangent to the actual line, an integrated rate law measures the actual line itself. Each integrated rate law is derived also using calculus, and your understanding of its derivation is not important.

In every integrated rate law equation we need solve for k, the rate constant. From that, we can solve any other question. The most important thing when using integrated rate laws is to first determine the order of the reaction, so you know which integrated equation to use. The reason is that each reaction order uses an entirely different integrated equation, and the hardest part to this topic is memorizing them! However, once you know which equations belong to which order reactions you are set. From that point, all you have to do is plug in the numbers given, and you should be home free.

The following are the equations that you must use for the different reaction orders:

0th order reaction: $[A]_o - [A]_t = kt$

1st order reaction: $\ln\frac{[A]t}{[A]_o} = -kt$

2nd order reaction: $\frac{1}{[A]_t} - \frac{1}{[A]_o} = -kt$

3rd order reaction: $\frac{1}{2[A]_t^2} - \frac{1}{2[A]_o^2} = -kt$

All of these use the following components:

$[A]_o$ = initial concentration
$[A]_t$ = concentration at time t
k = rate constant
t = time

Example: In the 1st order decomposition of pentane (C_5H_{12}) to pentene (C_5H_{10}), how long will it take for 80 % of pentane to react? Use the information provided in the table.

Time (min)	$[C_5H_{12}]$ (M)
0	4
5	2
10	1
15	0.5

Step 1: First find the reaction constant

This is given as a first order reaction, so we need to use the integrated rate equation for first order reactions. Simply rearrange the equation to solve for k, and use t_0 = 0 and t = 15 min. (Hint: use points that are the furthest apart to obtain the most accurate results).

$$\ln\frac{[C_5H_{12}]}{[C_5H_{12}]_o} = -kt$$

$$k = -\frac{\ln\frac{[C_5H_{12}]}{[C_5H_{12}]_o}}{t}$$

$$k = -\frac{\ln\frac{[0.5]}{[4]_o}}{15}$$

$$k = 0.139$$

Step 2: Now we need to know how much time it takes to use 80 % of the pentane

To solve this, we will go back to our 1st order reaction equation and solve for t. The tricky part here is to determine what values to use for the concentration of pentane.

When you think about it, the fraction $\frac{[C_5H_{12}]}{[C_5H_{12}]_o}$

refers to the amount of material remaining. Therefore, if 80 % was used, 20 % is remaining. So we can use 0.20 for this fraction since we are solving for how long it takes to get down to 20 % left.

$$\ln\frac{[C_5H_{12}]}{[C_5H_{12}]_o} = -kt$$

$$t = -\frac{\ln\frac{[C_5H_{12}]}{[C_5H_{12}]_o}}{k}$$

$$t = -\frac{\ln(0.20)}{0.139}$$

$$t = 11.6\ \text{min}$$

4. HALF LIFE

Half-life (symbol $t_{1/2}$) refers to the amount of time it takes for half of the starting material to disappear. Half-life calculations are important for two reasons. In 1st order reactions, once we have $t_{1/2}$ we can easily get k. Also, we can date old objects using the half-life of a radioactive material. The material that is most typically used in dating is Carbon-14.

The equations for half-life corresponding to the order of the reaction:

0th order: $t_{1/2} = \frac{[A]_0}{2k}$

1st order: $t_{1/2} = \frac{0.693}{k}$

2nd order: $t_{1/2} = \frac{1}{k[A]_0}$

3rd order: $t_{1/2} = \frac{3}{2k[A]_o^2}$

4a. Using half-life to solve for k or vice versa

One of the most important calculations involving half-life is converting between half-life and k. This is easiest for first-order reactions, since the relationship for the 1st order reaction between $t_{1/2}$ and k shows we can solve for one given the other. This is important when trying to work some problems dealing with integrated rate laws:

$$t_{1/2} = \frac{0.693}{k} \quad \text{and} \quad k = \frac{0.693}{t_{1/2}}$$

4b. Dating old objects

Dating old objects is a typical question that is asked under the topic of a first-order half-life problem. Typically the problem will give you the half-life of the radioactive dating material, the initial decay rate of the dating material and the decay rate of the object in question. Using the integrated rate law, you can solve for t to give the age of the object.

Example: An old artifact was found and radiocarbon dated. The half-life for carbon-14 is 5,730 years. The decay rate for carbon-14 in nature is 15 $dis.min^{-1}$ but was only 5 $dis.min^{-1}$ in the artifact. How old is the artifact?

Step 1: Use half-life given in the problem to solve for k:

$$k = \frac{0.693}{t_{1/2}} = \frac{0.693}{5730\ \text{yr}} = 1.21 \times 10^{-4}\ \text{yr}^{-1}$$

Step 2: Write the integrated rate law for 1st order reaction

$$\ln\frac{[A]_T}{[A]_o} = -kt$$

Step 3: Recognize that $[A]_T$ = 5 dis.min^{-1} and $[A]_0$ = 15 dis.min^{-1} and then solve for t:

$$t = -\frac{\ln([A]_T/[A]_0)}{k}$$

$$t = -\frac{\ln(5/15)}{1.21\times10^{-4}\ \text{yr}^{-1}}$$

$$t = 9{,}079 \text{ years}$$

SUMMARY

Kinetics is the study of rates of chemical reactions. Usually the study of kinetics involves the use of calculus. However, in a general chemistry course you will not be responsible for any calculus. Make sure you remember how to obtain a kinetic rate equation by first finding the order of each reactant and then using one trial to solve for k. In reaction mechanisms, be sure you understand that the slow step is the rate determining step. Be aware that integral rate laws are the line in a graph as opposed to the tangent line drawn to the original line in the graph. Lastly, know that the overall order of the reaction determines which integrated rate law equation and which half-life equation to use.

CHAPTER 10

LEWIS STRUCTURES AND VSEPR THEORY

OVERVIEW

Up until now we have always been given chemical formulas but never had to draw the structures ourselves. In this topic of introductory chemistry you will be responsible for drawing structures on your own. There is a lot involved in drawing chemical structures, such as determining how many paired and lone paired electrons there are; the types of bonds between atoms; the orientation of atoms around a central atom, and specific characteristics about the overall molecule. This chapter addresses all of the important things that you will need to know about drawing your own molecules and being able to answer questions about the molecules that you drew. First we will learn how to draw Lewis structures based on the number of valence electrons, and important attributes about Lewis structures. Then we will talk about the Valence Shell Electron Pair Repulsion (VSEPR) theory and how to determine the orientation of the atoms. Lastly we will address molecular orbitals and bonding and anti-bonding electrons.

1. LEWIS STRUCTURES

1a. Determining number of valence electrons

The most important part of drawing a Lewis structure is to first be able to determine the number of valence electrons that the molecule has. Valence electrons are the electrons in the outermost energy level. Each element in the molecule brings in its own specific number of valence electrons; meaning the overall molecule will have as many valence electrons as the sum of all of the elements' valence electrons. The way to determine how many valence electrons each element has is by simply looking at where the element is on the periodic table. The periodic table is organized in such a way so that all of the elements in the same column (vertical direction) have the same number of valence electrons. Below is a summary of the number of valence electrons for each column on the periodic table.

1	**2**											**3**	**4**	**5**	**6**	**7**	**8**
1	2	3	4	5	6	7	8	9	10	11	12	13	14	15	16	17	18
1 H Hydrogen 1.0079																	2 He Helium 4.0026
3 Li Lithium 6.941	4 Be Beryllium 9.0122											5 B Boron 10.811	6 C Carbon 12.011	7 N Nitrogen 14.0067	8 O Oxygen 15.9994	9 F Fluorine 18.9984	10 Ne Neon 20.1797
11 Na Sodium 22.9898	12 Mg Magnesium 24.3050											13 Al Aluminum 26.9815	14 Si Silicon 28.0855	15 P Phosphorus 30.9738	16 S Sulfur 32.066	17 Cl Chlorine 35.4527	18 Ar Argon 39.948
19 K Potassium 39.0983	20 Ca Calcium 40.078	21 Sc Scandium 44.9559	22 Ti Titanium 47.88	23 V Vanadium 50.9415	24 Cr Chromium 51.9961	25 Mn Manganese 54.9380	26 Fe Iron 55.847	27 Co Cobalt 58.9332	28 Ni Nickel 58.693	29 Cu Copper 63.546	30 Zn Zinc 65.39	31 Ga Gallium 69.723	32 Ge Germanium 72.61	33 As Arsenic 74.9216	34 Se Selenium 78.96	35 Br Bromine 79.904	36 Kr Krypton 83.80
37 Rb Rubidium 85.4678	38 Sr Strontium 87.62	39 Y Yttrium 88.9059	40 Zr Zirconium 91.224	41 Nb Niobium 92.9064	42 Mo Molybdenum 95.94	43 Tc Technetium (98)	44 Ru Ruthenium 101.07	45 Rh Rhodium 102.9055	46 Pd Palladium 106.42	47 Ag Silver 107.8682	48 Cd Cadmium 112.411	49 In Indium 114.82	50 Sn Tin 118.710	51 Sb Antimony 121.757	52 Te Tellurium 127.60	53 I Iodine 126.9045	54 Xe Xenon 131.29
55 Cs Cesium 132.9054	56 Ba Barium 137.327	57 La Lanthanum 138.9055	72 Hf Hafnium 178.49	73 Ta Tantalum 180.9479	74 W Tungsten 183.85	75 Re Rhenium 186.207	76 Os Osmium 190.2	77 Ir Iridium 192.22	78 Pt Platinum 195.08	79 Au Gold 196.9665	80 Hg Mercury 200.59	81 Tl Thallium 204.3833	82 Pb Lead 207.2	83 Bi Bismuth 208.9804	84 Po Polonium (209)	85 At Astatine (210)	86 Rn Radon (222)
87 Fr Francium (223)	88 Ra Radium (226)	89 Ac Actinium (227)	104 Rf Rutherfordium (265)	105 Db Dubnium (268)	106 Sg Seaborgium (271)	107 Bh Bohrium (270)	108 Hs Hassium (277)	109 Mt Meitnerium (276)	110 Ds Darmstadtium (281)	111 Rg Roentgenium (280)	112 Cn Copernicium (285)	113 Uut Ununtrium	114 Fl Flerovium (289)	115 Uup Ununpentium	116 Lv Livermorium (293)	117 Uus Ununseptium	118 Uuo Ununoctium

58 Ce Cerium 140.115	59 Pr Praseodymium 140.9076	60 Nd Neodymium 144.24	61 Pm Promethium (145)	62 Sm Samarium 150.36	63 Eu Europium 151.965	64 Gd Gadolinium 157.25	65 Tb Terbium 158.9253	66 Dy Dysprosium 162.50	67 Ho Holmium 164.9303	68 Er Erbium 167.26	69 Tm Thulium 168.9342	70 Yb Ytterbium 173.04	71 Lu Lutetium 174.967
90 Th Thorium 232.0381	91 Pa Protactinium 231.0359	92 U Uranium 238.0289	93 Np Neptunium (237)	94 Pu Plutonium (244)	95 Am Americium (243)	96 Cm Curium (247)	97 Bk Berkelium (247)	98 Cf Californium (251)	99 Es Einsteinium (252)	100 Fm Fermium (257)	101 Md Mendelevium (258)	102 No Nobelium (259)	103 Lr Lawrencium (262)

** Notice that group **1** elements have **1 valence electron**, group **2** element have **2**; then group 1**3** have **3**, 1**4** have **4**, etc. So if ignore the 1 in front of each number, then the group number gives you directly the valence electrons!

1b. A systematic approach to drawing Lewis Structures

Drawing Lewis structures follows the same steps every time. As long as you follow the steps outlined below, you will be able to draw a Lewis Structure for any molecule given.

Step 1: Add up the total amount of valence electrons from each element in the molecule

Step 2: Draw a 'bare-bone' structure of the molecule with an element in the center and one bond connecting all the other elements to the central element

Step 3: Counting each bond as having 2 electrons, add up the total amount of electrons used in the bare-bone structure, then subtract this from the total number of valance electrons

Step 4: With the remaining electrons, fill in all of the outer elements that are bonded to the central element with electron pairs until they all have 8 electrons. This is because all elements need to have an octet to be stable. (Remember, you are really only adding 6 electrons to each element because they each already have 2 in the bond drawn between that element and the central element). Finally, remember here that the one key exception is hydrogen, which can never hold more than 2 electrons.

Step 5 (if necessary): In the event that there are too many or too little electrons:

Too little: If there are not enough valence electrons to give each element – usually the central element - a stable octet you must share an extra pair of electrons with an outer element and the central element, to make sure they each have a stable octet. However, note that exceptions here are the elements Be, B or Al, which can be stable with less than 8 electrons.

Too many: If every element has a stable octet but you still have more electrons left over from the original sum of valence electrons you must put the remainder electrons on the central element. As long as the central element occurs in Period 3 or higher, this is OK, as those elements can take more than 8 if needed.

Example: Draw the Lewis Structure for CF_4

Step 1: Add up the total amount of valence electrons from each element in the molecule

C: 1 × 4 = 4
F: 4 × 7 = 28
Total Valence in molecule = 32

Step 2: Draw a bare-bone structure of the molecule with carbon as the central atom and surrounded by 4 fluorines, each with a single bond to the central carbon:

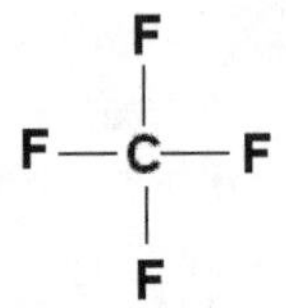

```
      F
      |
  F — C — F
      |
      F
```

Step 3: Add up the total amount of electrons that we used in the bare-bone structure and subtract this amount from the total number of valance electrons from step 1. Since we drew 4 bonds, each with 2 electrons, this is 8 total:

32 – 8 = 24

Step 4: With the amount of electrons that you have left (24 in this case), fill in all of the elements that are bonded to the central element with electron pairs until they each have 8 electrons.

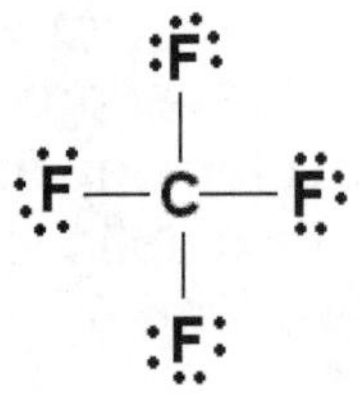

```
       ..
      :F:
       |
  :F̤ — C — F̤:
   ..       ..
       |
      :F:
       ..
```

Step 5: Verify you used the correct amount of electrons and that each element has a stable octet

24 – 24 = 0

2. RESONANCE

Electrons that make bonds in molecules can be thought of as a 'sea of mobile particles', meaning that they are constantly moving. The word resonance explains how electrons resonate back and forth from element to element in a molecule. Not all electrons are capable of resonating; only electrons that are part of a double or triple bond or electrons that are lone pairs. Molecules that have lone pairs and/or double and triple bonds are therefore capable have having resonance structures. In a resonance structure, it is the same molecule with the same orientation of electrons but electrons have been moved. A typical example of a resonance structure would be that of a molecule with one structure having lone pairs around oxygen and the other structure taking one of the lone pairs and making it into a double bond.

The following is an example of resonance in ozone (O_3):

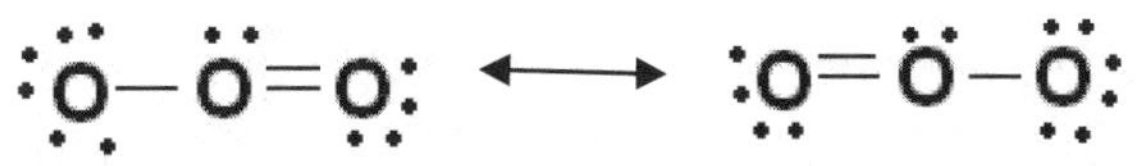

:O–O=O: ⟷ :O=O–O:

**Note – all resonance structures have the exact same orientation of the elements, the only difference between each structure is the movement of pairs of electrons. Resonance structures are different than isomers.
**Note – all resonance structures are illustrated using a double headed arrow

3. FORMAL CHARGE

Now that we know how to draw Lewis Structures, how do we know which combination of lone pairs and double bonds is the best to draw? The answer to this question deals with something known as formal charge. Formal charge serves as a means of determining which structure is the most stable. The rule for determining the best structure is to find the one with the smallest formal charges and the least amount of net formal charge. In order to find the one with the least amount of formal charge, we first have to know how to calculate formal charge.

Calculating formal charge is very simple. The formula that I use to calculate formal charge on each element is: Formal charge = # Valence electrons – dots – dashes. In other words, for each element take its number of valence electrons and subtract from that number how many dots are around the element and how many dashes are around the element. This is a bulletproof technique that works every time.

**Note to make sure you count <u>every</u> dash in double and triple bonds and do not just count them as 1.

Example: Calculate the formal charge for each element in the following form of O_3:

$:\ddot{\underset{\cdot\cdot}{O}}-\ddot{O}=\underset{\cdot\cdot}{O}:$

Formal charge on 1st oxygen: 6 valence electrons, 6 dots, 1 dash: 6 – 6 – 1 = -1
Formal charge on 2nd oxygen: 6 valence electrons, 2 dots, 3 dashes: 6 – 2 – 3 = +1
Formal charge on 3rd oxygen: 6 valence electrons, 4 dots, 2 dashes: 6 – 4 – 2 = 0

So we can write: $:\ddot{\underset{\cdot\cdot}{O}}-\ddot{O}=\underset{\cdot\cdot}{O}:$

-1 +1 0

Notice how this previous example is a better Lewis Structure for the molecule O_3 than:

Formal charges: $:\ddot{\underset{\cdot\cdot}{O}}-\ddot{O}-\ddot{\underset{\cdot\cdot}{O}}:$

-1 +2 -1

The reason for this is because this form of O_3 has formal charges of -1, +2, and -1 respectively. Since this form has more formal charge, it is not as stable.

4. POLARITY

In a separate chapter we talked about electronegativity and how it is an element's ability to attract the electrons in a covalent bond. Each element on the periodic table has a different electronegativity, meaning each element has a slightly different pull on electrons when it makes a bond. I like to think of this as a game of tug of war. A covalent bond is a bond between two non-metals in which the electrons are shared between the two elements. However, it is not always an equal sharing. When an element with a higher electronegativity is bonded to an element with lower electronegativty, there will be an unequal sharing of the electrons. This would result in the electrons being pulled more closely to the element with the higher electronegativity.

As we know electrons are negatively charged, causing a slightly negative charge on that end of the bond. The lack of electrons on the other side of the bond results in a slightly positive charge. Having a bond with a slightly positive end and a slightly negative end creates a polarized, or polar, bond. A bond between two atoms of the same elements results in an equal sharing of the electrons and is non-polar.

Now that we know the difference between polar and non-polar bonds, let's look at what is meant by a polar or non-polar molecule. The key to determining the polarity of a molecule is symmetry. Symmetric molecules with the same elements on all sides results in a non-polar shape. Going back to the tug of war analogy, having a symmetric molecule with the same element on all sides results in equal pulls in the opposite directions which cancel each other out. Keep in mind that there can still be polar bonds within the non-polar molecule, but the symmetry causes the individual polarities to cancel out, resulting in a net non-polar molecule.

To illustrate the polarity in a bond we use an arrow notation in which the arrow starts at the positive end and points towards the negative end of the bond. A good way to remember this is that the tail of the arrow looks like a plus sign.

Example illustrating polarity:

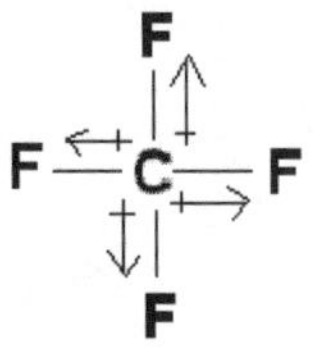

Note: The actual shape of CF_4 is tetrahedral (see the VSEPR section). Even though each C-F bond is polar, the symmetric shape of CF_4 causes it overall to be non-polar.

5. HYBRIDIZATION

Have you ever wondered the reason why elements are capable of forming four bonds? The way that this happens is through a process called hybridization. Hybridization occurs when the orbital in the s subshell combines with one or more of the the orbitals in the p subshell to form hybrid orbitals. The number of bonds the particular element can form will determine the hybrid orbitals needed when it forms a compound. For an element with 2 bonds, the s orbital combines with one p orbital, resulting in two sp hybrid orbitals. In a molecule where 3 bonds are needed, an s orbital combines with 2 p orbitals resulting to form three sp^2 hybrid orbitals. Finally, 4 bonds require an s orbital to hybridise with 3 p orbitals resulting in a total of four sp^3 hybrid orbitals.

As mentioned previously, there are some elements that go beyond 4 bonds and have what are called extended octets. This is only possible with elements that have a d subshell. The reason is because there are only 3 p orbitals, so any additional bonding after 4 bonds would require starting to tap into the d orbitals. Therefore, to have 5 bonds, an element will require 1 s orbital, 3 p orbitals, and 1 d orbital to create five sp^3d hybrids. Lastly, to have 6 bonds, 1 s orbital, 3 p orbitals, and 2 d orbitals are needed, resulting in six sp^3d^2 hybrid orbitals.

The following table is a summary of the hybrid orbitals formed when elements bond:

# of Bonds	Hybrid Orbitals
2	sp
3	sp^2
4	sp^3
5	sp^3d
6	sp^3d^2

*Note that the number of bonds is equal to the number of orbitals present in the hybrid. Notice how 2 bonds requires 2 orbitals, 3 bonds require 3 orbitals, etc.

6. VSEPR THEORY

Valence Shell Electron Pair Repulsion, which is also known as VSEPR, is a theory that is used to predict the shape and orientation of bonds and shapes of molecules. The way that we determine shape is based solely on the central element in the molecule. The idea behind the VSEPR theory is that bonds and lone pairs are negatively charged; so they repel each other and want to be as far away from each other as possible. So certain molecular shapes arise to alleviate the repulsion between electrons.

6a. Determining Electron Domains

Before we get into the shapes of molecules we must first discuss how to be able to identify the shape based on the central element. To do this, instead of counting how many bonds the central element has, we count electron domains. An electron domain is simply an area around an element where electrons exist. It is important to note that double/triple bonds count as ONE electron domain, and also lone pairs count as one electron domain. Once you determine the number of electron domains around the central element you will be able to determine the shape of any molecule.

Example of counting electron domains:

The following molecule has 5 electron domains. My best advice for counting electron domains is to circle them as you count them to help you visualize them and help you avoid either missing one or counting one twice.

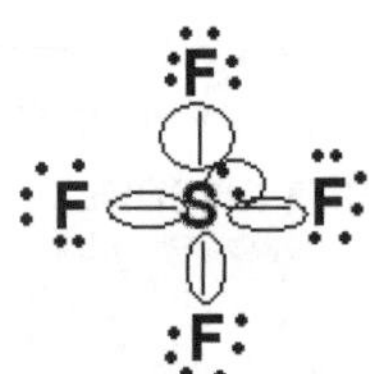

Once you have determined the correct number of electron domains, you are ready to determine the shape of the molecule. There are two different types of geometry that we use to identify molecular shapes. The electronic geometry determines the shape based solely on the number of electron domains. The molecular geometry determines a name based on the actual bonds between atoms in the molecule. The next two sections summarize how to figure out shapes for molecular or electronic geometries.

6b. Electronic Geometry

Again, the electronic geometry determines the shape based solely on the number of electron domains. The following table summarizes the electronic geometries for the different molecules based on the number of electron domains.

Electron Domains	Electronic Geometry
2	Linear
3	Trigonal Planar
4	Tetrahedral
5	Trigonal Bipyramidal
6	Octahedral

6c. Molecular Geometry

The molecular geometry is a little more specific than the electronic geometry. For the molecular geometry, you must take into account how many domains are actual bonds between elements, and how many are from lone pairs. For each additional set of lone pairs the molecular geometry changes. The following table summarizes the molecular geometries based on how many sets of lone pairs there are in the electron domain.

Electron Domain	Sets of Unpaired	Molcular Geometry
2	0	Linear
3	0	Trigonal Planar
	1	Bent
4	0	Tetrahedral
	1	Trigonal Pyramidal
	2	Bent
5	0	Trigonal Bipyramidal
	1	Seesaw
	2	T Shape
	3	Linear
6	0	Octahedral
	1	Square Pyramidal
	2	Square Planar

**Note that the molecular geometry is the same as the electronic geometry when there are no sets of unpaired electrons.

Example: What are the electronic and molecular geometries of H_2O?

Step 1: Draw the Lewis structure (refer to the steps in the Lewis Structure section):

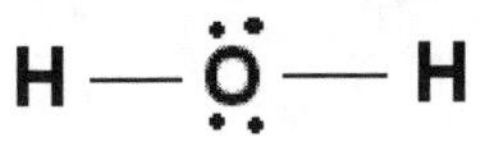

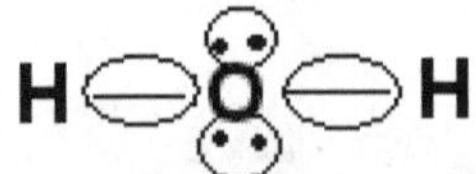

Step 2: Count the electron domains

Electron Domains = 4

Step 3: Determine Electronic geometry:

Tetrahedral (memorize from table)

Step 4: Determine molecular geometry:

Bent (4 electron domains, 2 unpaired sets of electrons)

7. BOND ANGLES

You are probably sitting there thinking to yourself, "Okay great: I can memorize a table of names but I have no idea what it means or why I am memorizing it." Remember that the reason why the molecules form the shapes as indicated in the tables above is because the satellite elements want to spread away from each other as much as possible. In order for this to happen, there is a specific bond angle that forms between the elements. The following table describes the bond angles that you will need to memorize for the respective electron domains. (Technically the bond angles are modified slightly for each successive set of unpaired electrons. However, the good news is that you will be responsible for only these numbers in an introductory level course.)

Electron Domains	Bond Angle
2	180°
3	120°
4	109.5°
5	$90^{\circ}, 120^{\circ}$
6	90°

Example: What is the H-O-H bond angle in H_2O?

Answer: since H_2O has a tetrahedral geometry, the H-O-H bond is 109.5°

8. SIGMA BONDS VS. PI BONDS

Instructors love to give you a molecule and ask you how many sigma and pi bonds are in the molecule. These questions are extremely easy, but you first have to know how to identify sigma and pi bonds.

Sigma Bond

A sigma bond is a bond between hybrid orbitals. This is the standard type of covalent bond and is always found in a single bond.

Pi Bond

A pi bond is best explained as a bond between unhybridised p orbital electrons. You should think of pi bonds as additional bonds that form after an already existing sigma bond. Pi bonds are found in double and triple bonds. A double bond has 1 sigma and then 1 pi bond. A triple bond has 1 sigma bond and then 2 pi bonds.

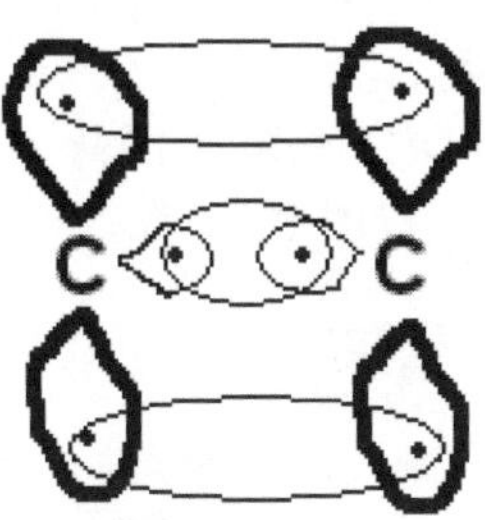

(This would be an example of a double bond)

Example: How many sigma and pi bonds are in the following molecule?

```
   H   :O:  H   H
   |   ||   |   |
H— C — C — C == C — C ≡ C — H
   |
   H
```

Answer: 12 Sigma Bonds; 4 Pi Bonds

Explanation: The easiest way to determine the number of sigma bonds is to simply count every bond once. You will notice there are 12 bonds connecting all of the elements together. To count the pi bonds, simply count all of the extra bonds that are present in addition to the single bonds. There are two double bonds with 1 pi bond each. There is also a triple bond with 2 pi bonds. So this is a total of 4 pi bonds.

9. MOLECULAR ORBITALS

VSEPR is good for explaining molecular structure, however, there are a few things that cannot be explained using VSEPR theory. As a result, there is another way of looking at molecules which focuses on the entire molecule instead of just individual atoms. With this new way of analyzing a molecule, we can determine properties such as magnetism and the bond order of molecules.

9a. Forming molecular orbitals

Molecular orbitals, once formed, are filled similar to the way atomic orbitals are filled. However, when filling molecular orbitals you must be aware of bonding vs. anti-bonding orbitals. Anti-bonding orbitals are denoted with an asterisk, and this distinction will become particularly important when we discuss bond order in section 9c.

When filling molecular orbitals, you must first count the number of valence electrons that are present in the molecule. Once you have determined the number of electrons, you are then ready to start filling the orbitals. The order in which the orbitals fill is sigma, sigma anti-bonding, pi, sigma, pi anti-bonding, sigma anti-bonding. All of the sigmas have one orbital and all of the pi's have two orbitals. This information is summarized in the following diagram below:

σ^* —
π^*— —
σ —
π — —
σ^* —
σ —

The way that I suggest to memorize the following information is to make a song or a beat to the sequence, 112121. You will notice that this sequence is the order of orbitals in the 'staircase' above. Once you remember this, the rest of the information will fit in relatively easily because for every 1 line you will know we have a sigma orbital, and for the 2 lines you will know we have a pi orbital. As far as the anti-bonding is concerned just remember "2nd and last 2". The second molecular orbital, and the last 2 molecular orbitals are anti-bonding.

Example: give the electron arrangement in the molecular orbitals of O_2:

Step 1: Determine number of valence electrons

O: 6 × 2 = 12

Step 2: Recall sequence of molecular orbital filling

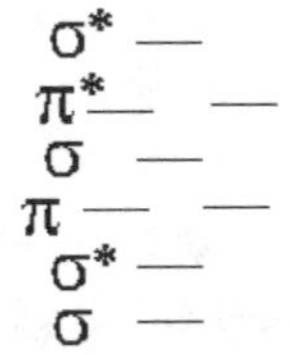

σ^* —
π^* — —
σ —
π — —
σ^* —
σ —

Step 3: Fill in the 12 valance electrons starting at the bottom. Each orbital must have 2 electrons before continuing to the next level. In the pi orbitals, you must place an electron in each orbital before you start to double up.

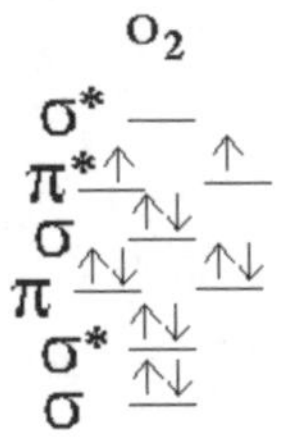

O_2
σ^* —
π^* ↑ ↑
σ ↑↓
π ↑↓ ↑↓
σ^* ↑↓
σ ↑↓

9b. Magnetism

For magnetism you will have to be familiar with two important terms; diamagnetic and paramagnetic. **Diamagnetic** refers to a molecule with no lone pairs of electrons in the molecular orbital configuration. Molecules that are diamagnetic actually repel a magnet. **Paramagnetic** refers to a molecule with unpaired electrons in the molecular orbital configuration. Molecules that are paramagnetic would be attracted to a magnet.

Example: Is oxygen diamagnetic or paramagnetic?

Answer: Oxygen gas, O_2, as seen in the example above is paramagnetic.

Explanation: You can see this by noticing there are 2 unpaired electrons in the molecular orbital configuration, so it is paramagnetic.

9c. Bond Order

Bond order is the way in which we can determine which bonds will be single , double or triple. The way that this is determined is by using the equation for bond order:

$$BO = \frac{(\#\, e^- \text{ in Bonding}) - (\#\, e^- \text{ Anti} - \text{bonding})}{2}$$

To calculate bond order for any bond simply count the number of electrons that are in bonding molecular orbitals and subtract the number of electrons that are in anti-bonding molecular orbitals. Then take that quantity and divide it by 2. You should get an answer of 1, 2, or 3. A bond order of 1 tells you that it is a single bond. A bond order of 2 tells you that the bond is a double bond, and a bond order of 3 tells you that the bond is a triple bond.

**Note that O_2 always has a bond order of 2 and N_2 always has a bond order of 3. These are two extremely popular questions that you will be asked on a test. Therefore, memorize O_2 is double bonded and N_2 is triple bonded.

Example: Determine the bond order of N_2

Step 1: Determine number of valence electrons

N: 5 × 2 = 10

Step 2: Recall sequence of molecular orbital filling

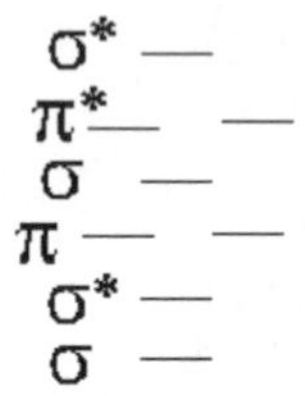

σ* —
π* — —
σ —
π — —
σ* —
σ —

Step 3: Fill in the 10 valance electrons starting at the bottom. Each orbital must have 2 electrons before continuing to the next level.

N_2

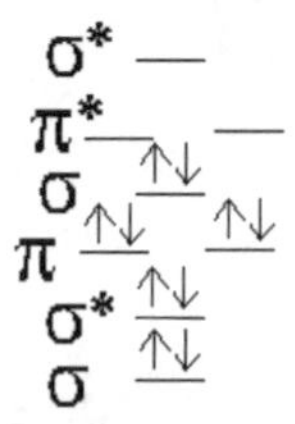

σ* —
π* — —
σ ↑↓
π ↑↓ ↑↓
σ* ↑↓
σ ↑↓

Step 4: Write Bond Order equation

$$BO = \frac{(\#\, e^- \text{ in Bonding}) - (\#\, e^- \text{ Anti} - \text{bonding})}{2}$$

Step 5: Plug in information from configuration in step 3

$$BO = \frac{(8)-(2)}{2}$$

Step 6: Solve

BO = 3 (Therefore N_2 is triple bonded)

SUMMARY

Electron Domain	Hybridization	Bond Angle	Electronic Geometry	Lone Pairs	Molecular Geometry	Polarity*
2	sp	180	Linear	0	Linear	Non-Polar
3	sp^2	120	Trigonal Planar	0	Trigonal Planar	Non-Polar
				1	Bent	Polar
4	sp^3	109.5	Tetrahedral	0	Tetrahedral	Non-Polar
				1	Trigonal Pyramidal	Polar
				2	Bent	Polar
5	sp^3d	90,120	Trigonal Bipyramidal	0	Trigonal Bipyramidal	Non-Polar
				1	Seesaw	Polar
				2	T Shape	Polar
				3	Linear	Non-Polar
6	sp^3d^2	90	Octahedral	0	Octahedral	Non-Polar
				1	Square Pyramidal	Polar
				2	Square Planar	Non-Polar

*Assuming every outer atom is of the same element

CHAPTER 11

NOMENCLATURE

OVERVIEW:

The word nomenclature is a very scary looking word that horrifies chemistry students all the time. The word nomenclature very simply means naming. This topic refers to the system that is set in place for naming chemical compounds. From now on, when you see the word nomenclature there is no longer a reason to be scared. This chapter will talk about the ways to name chemical compounds. The easiest way to break this material down is to organize it in your mind based on the type of bond that exists between the elements. The first thing that you want to do is identify what type of compound you are dealing with, and then go about naming it. The sections of this chapter are broken down into the different types of compounds and the rules for naming each particular type.

1. IONIC BOND NOMENCLATURE

An ionic bond is a bond between a metal and a non-metal. Remember a metal is an element on the left side of the staircase and a non-metal is an element on the right side of the staircase. When a metal and a non-metal unite, they bond in ratio to their charges. Meaning, if a +1 metal bonds with a -1 non-metal, this will be a 1 to 1 ratio in which only one element of each will bond together. However, in a case of a +2 metal bonding with a -1 non-metal, this is a 2 to 1 ratio in which there will be 2 non-metals for every one metal. The method that I use to explain this is called the criss-cross method. The criss-cross method is explained in the following example:

Na^{+1} Cl^{-1} forming NaCl

Ba^{+2} Cl^{-1} forming $BaCl_2$

As you can see the charge on the first element becomes the subscript on the second element and the charge on the second element becomes the subscript on the first element. The reason for this is simple. In the case of $BaCl_2$, Ba wants to lose two electrons to obtain a stable octet but Cl only wants to gain one electron. Therefore Ba will not be satisfied with only 1 Cl and will need 2 of them to get the job done.

Now that we are more familiar with ionic bonds, let's discuss how to name them. Since we know the charges on the metals and the non-metals, the formula for ionic bound molecules is already understood. This means we do not need to specify the amount of each element in the name for ionic compounds. This is very important, because this is not the case for other nomenclature.

To name a molecule with an ionic bond the rule is:

Name of the metal + root name of the non-metal + ide

Examples of naming ionic compounds:

NaCl = Sodium + Chlor + ide = Sodium Chloride

KF = Potassium + Fluor + ide = Pottassium Fluoride

$BaCl_2$ = Barium + Chlor + ide = Barium Chloride

Again, notice that even though there are 2 Chlorides in $BaCl_2$, we do not need to specify that in the name for ionic molecules. It is already implied due to the fact that the reader is expected to understand that it is a 2 to 1 ratio due to the charges on each element. (When a transitional metal is present, the rule is a little different and is discussed later in this chapter)

2. COVALENT BOND NOMENCLATURE

A covalent bond is a bond between two non-metals. Therefore a covalent bond consists of two elements that are both on the right side of the staircase. When naming compounds with a covalent bond, unlike ionic compounds, it is important to specify the number of each element present in the compound. When naming compounds, the subscripts did not matter. However, they do matter with covalently bound molecules.

In order to identify the number of each element in the name of the compound, a set of prefixes is used. The prefixes are:

1 – mono	6 - hexa
2 – di	7 - hepta
3 – tri	8 - octa
4 – tetra	9 - nona
5 – penta	10 – deca

The prefixes are added to the name of the element depending on the number of that element present. One important thing to note is that if there is only 1 of the first element in the compound, you do not use the prefix mono. In other words, the name of any compound never starts with the word mono. Mono is only used when there is only 1 of the second element in the compound.

The formula that you should use for naming covalently bound molecules is:

(Prefix + name first element) + (Prefix + root name of second element + ide)

Examples of naming covalent bound molecules:

CO = Carbon + mon + ox + ide = Carbon Monoxide
*note that mono is not used for the first element
**note that it is not monoxide, since you always drop a vowel if two occur together

N_2O_5 = Di + nitrogen + pent + ox + ide = Dinitrogen Pentoxide

H_2O = Di + hydrogen + mon + ox + ide = Dihydrogen Monoxide (or water!)

CCl_4 = Carbon + tetra +chlor + ide = Carbon Tetrachloride

3. PRESENCE OF TRANSITION METALS

Whenever a transition metal is present, although it is still considered an ionic compound, the rules for naming are a little different. This is because transition metals can carry more than one charge; so it is necessary to specify which particular charge on the transition metal carries once in the compound. The way that the charge is determined is by looking at the other element in the compound. For example, in $FeCl_2$ there are two chlorines for every 1 iron, meaning that this had to be the Fe^{+2} ion. In $FeCl_3$ there are three chlorines for every 1 iron, meaning that this had to be the Fe^{+3} ion. Clearly we can see a difference between $FeCl_2$ and $FeCl_3$, but we must also be able to differentiate between the two of them in the name as well.

In order to differentiate between which charge on the transition metal was used, we used something called the stock system. The stock system is very simply the ionic bond nomenclature rule with the addition of roman numerals to denote the specific charge that was used. The Roman numeral corresponds to the charge and is written in between the name of the two elements.

The formula for ionic compounds with the presence of a transition metal:

Name of the metal +(roman numberal) + root name of the non-metal + ide

Examples of naming ionic bound molecules with a transition metal present:

$FeCl_2$ = Iron + (II) + chlor + ide = Iron (II) Chloride
$FeCl_3$ = Iron + (III) + chlor + ide = Iron (III) Chloride
CuO = Copper + (II) + ox + ide = Copper (II) Oxide

The stock system is very simple; just make sure you remember to use it. Anytime you see a transition metal in a compound in which an instructor is asking you to name, be sure you recognize the need to use the stock system and then use it correctly. Also, note that it is also used for main group metals such as Tin or Lea, which can also carry more than one charge.

3a: Using Latin names

I highly recommend using the stock system when transition metals are present. However, there is another way of naming transition metals that you should be able to recognize. Instead of using Roman numerals, we just use the Latin root of the transition metal along with the correct suffix ending. The suffix will imply whether it is the bigger or smaller of the two charges for that element. If we want to denote the smaller of the two charges, we end the metal name with –ous. If we want to denote the bigger of the two charges, we end the metal name with –ic. By using this convention, there is no need for Roman numerals: instead we incorporate the charge on the metal right into the name.

The formula for naming transition metal compounds using Latin suffixes:

Latin root + proper suffix + root name of second element + ide

Examples of naming compounds using Latin suffixes:

$FeCl_2$ = Ferr + ous + Chlor + ide = Ferrous Chloride
$FeCl_3$ = Ferr + ic + Chlor + ide = Ferric Chloride
CuO = Cupr + ic + ox + ide = Cupric Oxide

The most popular Latin roots that I recommend memorizing are:
Copper – Cupr
Gold – Aur
Iron – Ferr
Lead – Plumb
Tin – Stann

Interesting side notes
Latin roots are the reason why elements such as iron are written on the periodic table as Fe and not Ir. If you ever wondered why iron is represented as Fe, you now know it is because of its Latin root Ferr. Also, pipes in houses used to be made out of lead. The latin root for lead is Plumb. This is where the word "plumbing" comes from.

4. POLYATOMIC ANIONS

An anion, remember, is a negatively charged particle. A polyatomic anion is a group of covalently bound non-metals that have a net negative charge. Polyatomic anions surface a lot in chemistry and it is important to understand how to name them. Just like with transition metals, polyatomic anions exist in more than one form. For example there is an SO_3 and an SO_4. Therefore in the name of polyatomic anions we must specify exactly which one we talking about. In order to denote the molecule with the smaller number of the second element we end the word in –ite. In order to denote the molecule with the larger numberof the second element we end the word in –ate.

The following is a list of the most important names of polyatomic anions and their respective charges:

SO_3^{-2} - Sulfite NO_2^{-1} - Nitrite ClO^{-1} - Hypochlorite PO_4^{-3} - Phosphate
SO_4^{-2} - Sulfate NO_3^{-1} - Nitrate ClO_2^{-1} - Chlorite
ClO_3^{-1} - Chlorate
ClO_4^{-1}- Perchlorate
OH^{-1} - Hydroxide

Keep in mind that polyatomic anions act as a single unit. Even though they are composed of multiple elements, they still act as any other single anion non-metal.

Examples of naming compounds with polyatomic anions:

$CaSO_4$ – Calcium Sulfate
$Ba(OH)_2$ – Barium Hydroxide
Na_3PO_4 – Sodium Phosphate
$FeSO_4$ – Iron (II) Sulfate or Ferrous Sulfate

5. NAMING ACIDS

An acid is a molecule with hydrogen at the beginning. Whenever a molecule starts with hydrogen you must recognize it as an acid and then name it accordingly. There are two different distinctions that you should make when naming acids. You should ask yourself does this acid contain a polyatomic anion, or does it contain a single anion. Once you have answered this question, the rules for naming the acid are very simple.

5a. Naming an acid with a single anion

Whenever the hydrogen is accompanied by a single element, the name starts with the prefix hydro. It is then followed by the root name of the anion. The ending of the name always ends in ic. Lastly, the word acid follows the name.

The formula for single anion acids is:

Hydro + root name of anion + ic + acid

Examples of single anion acids:

HBr = Hydro + brom + ic Acid = Hydrobromic acid
HCl = Hydro + chlor + ic Acid = Hydrochloric acid
HI = Hydro + iod +ic Acid = Hydroiodic acid

5b. Naming an acid with a polyatomic anion

When a polyatomic anion is present in the acid, it is important to first identify the polyatomic anion. Once you have recognized the polyatomic anion and remembered its name you must use the following rule: If the ending of the polyatomic anion is 'ite', the name of the acid will end in ous. If the ending of the polyatomic anion is 'ate', the name of the acid will end in ic. For convenience, the following rule is summarized below:

$$\text{ite} \longrightarrow \text{ous}$$

$$\text{ate} \longrightarrow \text{ic}$$

The formula for naming an acid with a polyatomic anion is:

Root name of polyatomic anion + appropriate ending + Acid

Examples of naming an acid with a polyatomic anion:

HNO_2 = Nitr + ous + Acid = Nitrous Acid
HNO_3 = Nitr + ic + Acid = Nitric Acid
H_3PO_4 = Phosphor + ic + Acid = Phosphoric Acid

SUMMARY

Nomenclature is not as scary as it sounds. The most important thing you want to do is to first recognize what kind of molecule it is that you are asked name. Does it have ionic bonds or covalent bonds? Is there a transition metal or polyatomic anion present? Is there an H in front making it an acid? These are the kinds of questions you want to ask yourself when faced with the challenge of naming these inorganic molecules. Once you recognize what kind of molecule it is, try to recall the formulas that were illustrated in this chapter and it will be very straightforward from there.

From my experience, I have noticed that people tend to get most tripped up with the polyatomic anions. These polyatomic anions surface all throughout your study of chemistry and I highly recommend you commit them to memory. They will follow (or haunt!) you throughout your entire study of the field of chemistry and now would be the time to learn them. Make flash cards, think of a jazzy song or catch phrase - whatever will help you remember them. The ones that I have specifically pointed out in this chapter are by far the most popular ones and the most important for you to memorize. I definitely recommend at least being familiar with the other ones that are listed in textbooks and lecture manuals.

There is one more important point that I would like to address. Keep in mind that nomenclature questions can be asked both forwards and backwards. Not only might you be expected to be given a compound and have to name it, but you might be given a name and asked to write the compound. Therefore, be sure you understand where the name comes from and what each part of the name means in order to be able to come up with a compound that it is describing. A basic rule of thumb would be to ask yourself does my answer account for every aspect of this molecule, and will a reader be able to discern exactly what I am describing based on the name alone.

CHAPTER 12

ORGANIC CHEMISTRY

OVERVIEW

Organic chemistry is viewed as the scariest topic imaginable in the eyes of many introductory chemistry students. When people think of organic chemistry they usually think of huge carbon chains with hundreds of different side chains coming off the molecule in every direction with all sorts of weird shapes and structures. This chapter is designed to introduce you to the field of organic chemistry and walk you through a very systematic approach to learning the subject.

In this chapter we will discuss organic nomenclature, and learn how to name even the most complicated looking structures. We will also explore some common names of molecules and I will point out the most popular ones that usually appear on exams. We will also talk about functional groups, and we will see a very easy way of remembering the different functional groups. After that, we will talk about cyclic molecules and how to approach them. We will then conclude the chapter by addressing the different types of isomers and will also discuss organic reactions. (Organic reactions are usually not seen until the second semester of organic chemistry).

1. HYDROCARBON NOMENCLATURE

Most exam questions involving organic compounds come in the form of hydrocarbon nomenclature. The IUPAC Hydrocarbon nomenclature is the method of naming structures that contain only carbon and hydrogen atoms. Before we dive right in and start talking about how to name the complicated looking structures, we will start off with the basics. Once the basics are mastered, as you will see, you will be able to name even the ugliest of ugly molecules.

Every name is comprised of two components. The first component, the root, accounts for how many carbon atoms are in the chain. The second component, the suffix, accounts for what types of bonds are between the carbons. By putting these two components together, we can name molecules with enough information to be able to establish the molecule's structure with nothing but the name itself!

1a. Naming the Root

The first half of every name, the root, shows us how many carbons are in the chain. The numbering system that we use to denote the number of carbons is shown below:

1 Carbon = Meth	6 Carbon = Hex
2 Carbon = Eth	7 Carbon = Hept
3 Carbon = Prop	8 Carbon = Oct
4 Carbon = But	9 Carbon = Non
5 Carbon = Pent	10 Carbon = Dec

**Easy way of remembering these roots:

To remember the first four prefixes, which are the hardest to remember, I use the pneumonic device: **M**others **E**verywhere **P**lay **B**all

To remember the roots from 5 to 10, you will notice that they all fit with the regular geometric shapes that we have known since elementary school. For example, a 5 sided polygon is a **pent**agon and a molecule containing 5 carbons starts with pent. Also, a 6 sided polygon is a **hex**agon and a molecule containing 6 carbons starts with hex.

1b. Naming the suffix

The second half of every name, the suffix, accounts for what types of bonds exist between the carbon atoms. The three possibilities are single bonds, double bonds, and triple bonds. We group these three cases into three different distinct types of hydrocarbons.

Alkane:
An alkane is a hydrocarbon with single bonds only. If the molecule that we are trying to name is an alkane, the suffix will be –ane. Note that since the word alkane ends in –ane, so will the name of our molecule. The general formula for an alkane is C_nH_{2n+2} (where n is any integer from 1 upwards) because every carbon is bonded to 2 hydrogens and then the two carbons at the end of the chain each take an extra hydrogen as well.

An example of what an alkane looks like is shown below. You will notice that each carbon has 4 bonds because every carbon must make 4 bonds (totaling 8 valance electrons for a stable octet, as described in the chapter on Lewis structures).

```
    H   H   H
    |   |   |
H — C — C — C — H
    |   |   |
    H   H   H
```

This molecule has 3 carbons so the name starts with Prop-. It has all single bonds making it an alkane, so it ends with –ane. Putting that all together means the name for the above molecule is Propane.

Alkene:
An alkene is a hydrocarbon with a double bond. If the molecule that we are trying to name is an alkene, the suffix will be –ene. The formula for an alkene is C_nH_{2n}; meaning there is double the number of hydrogens as there are carbons. There are two less hydrogen atoms in an alkene compared with an alkane because of the presence of the double bond, since for each additional bond that is added to an organic molecule, 2 hydrogen atoms are no longer needed.

An example of what an alkene looks like is shown below:

```
H   H   H
|   |   |
C===C---C---H
|       |
H       H
```

This molecule has 3 carbons so the name starts with Prop-. It has a double bond making it an alkene and so it ends with –ene. Putting that all together, along with the fact that the double bond is on the first bond, means the name of this is 1-Propene.

Alkyne:
An alkyne is a hydrocarbon with a triple bond. If the molecule that we are trying to name is an alkyne, the suffix will be –yne. The formula for an alkene is C_nH_{2n-2}; meaning there is double the number of hydrogens as there are carbons, less two hydrogens. In other words, there are four less hydrogen atoms in an alkyne compared with an alkane because of the addition of 2 extra bonds making the triple bond over a snlge bond.

An example of what an alkyne looks like is shown below:

```
               H
               |
H---C≡C---C---H
               |
               H
```

This molecule has 3 carbons so the name starts with Prop-. It has a triple bond making it an alkyne and so it ends with –yne. Putting that all together, along with the fact that the triple bond is on the first bond, means the name of this is 1-Propyne.

1c. Naming Substituents

Now that we know how to name carbon chains, there is one other piece of information that we need in order to name complicated organic molecules. The chains of carbons that are not part of the longest carbon chain, but branch off from it, are called substituents.

There are three components to naming a substituent. The first component is a number in front of the name that designates which carbon in the longest chain the substituent group is attached to. For example, if the substituent is attached to the second carbon in the longest chain, than the number 2 precedes the name. It is important to number the carbons so that the substituent is preceded by the smallest number possible. The next component is the prefix that accounts for the number of carbons in the substituent. For your convenience, the prefixes are the same ones that are used for for single chain molecules. The last component of the name, the suffix, is –yl, and it is always the same for every substituent.

The following example shows how to name a substituent:

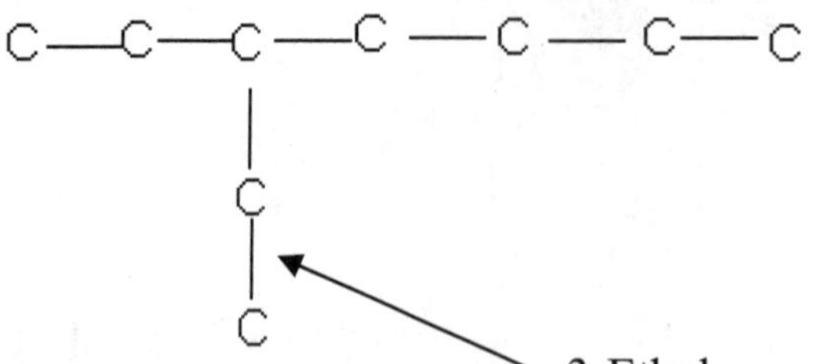

```
C—C—C—C—C—C—C
    |
    C
    |
    C
```

3-Ethyl

The reason why the name of the substituent is 3-ethyl is because it is located on the third carbon in the longest chain, it contains 2 carbons, and the suffix is always –yl.

1d. Steps for naming hydrocarbons

Now that we have seen the basics and understand how to name both straight chain hydrocarbons and substituents, let's put it all together in a very easy and systematic way. As long as you follow this systematic approach, you will be able to name any and all hydrocarbons.

1. Find the longest possible carbon chain and circle it
2. Number the carbons in the longest chain the direction that will give the smallest numbers for the substituents
3. Name the longest carbon chain and write it off to the side
4. Find and name any substituents and write them off to the side
5. Combine any like substituents
6. Write the substituents in alphabetical order
7. End with the name of the longest chain

Example: Name the following compound

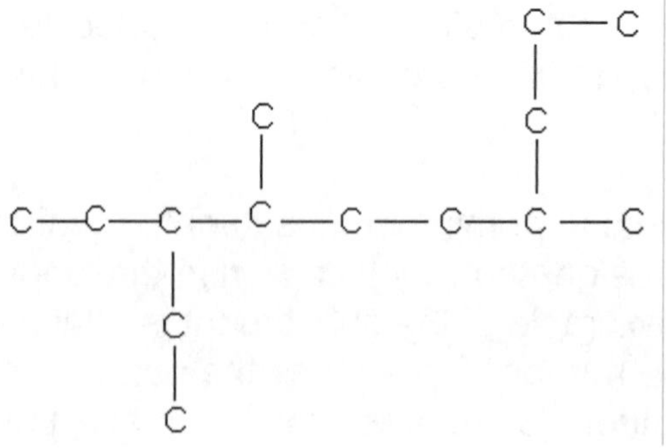

```
                  C—C
                  |
      C           C
      |           |
C—C—C—C—C—C—C—C
    |
    C
    |
    C
```

Step 1: Find the longest carbon chain and circle it:

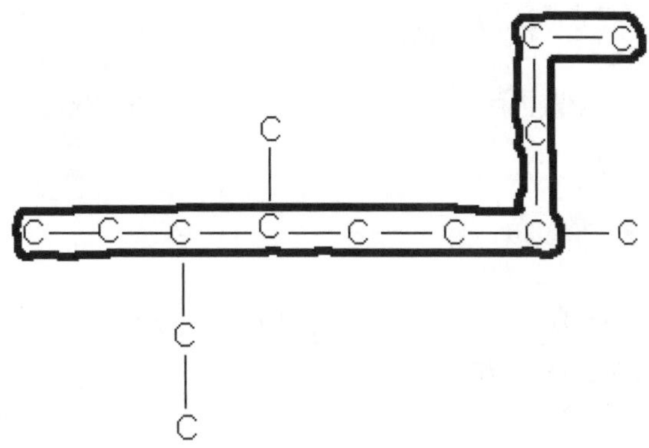

Step 2: Number the carbons in the longest chain the direction that will give the smallest numbers for the substituents:

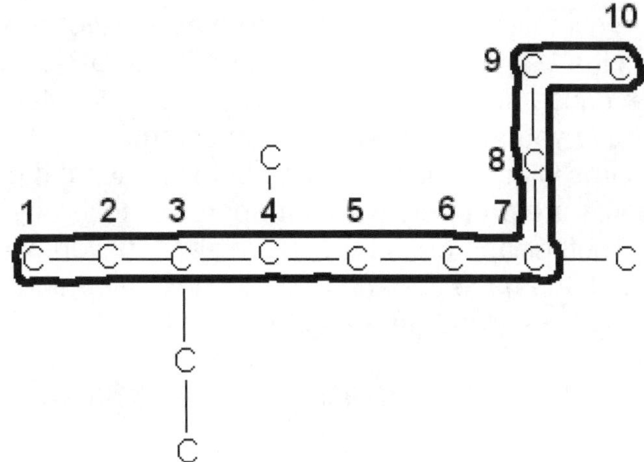

Step 3: Name the longest carbon chain and write it off to the side:

This has ten carbons, so it is a form of Decane

Step 4: Find and name any substituents and write them off to the side

3 – ethyl
4 – methyl
7 – methyl

Step 5: Combine any like substituents:

**Note – When you have two of the same substituent it becomes di-
When you have three of the same substituent it becomes tri-

3 – ethyl
4 – methyl
7 – methyl ⟶ 4,7 – dimethyl

Step 6: Alphabetize the substituents and write them in order:

3–ethyl–4,7–dimethyl

Step 7: Write the substituents in alphabetical order followed by the name of the longest chain:

3–ethyl–4,7–dimethyldecane

**Note – in the final answer, hyphens are used to separate numbers from letters and commas are used to separate numbers

2. COMMON NAMES

Some molecules have common names which serve as an alternate naming scheme. Most of the molecules that we name use the conventional IUPAC schemes; however, sometimes it is easier to use a common name for some molecules. In terms of an exam, most instructors will have you memorize dozens of common names. I highly suggest that you make flash cards with the name on one side and what the molecule looks like on the other. Flash cards are probably the best approach to wrapping your head around the common names. Another approach that I have seen to work well is using your creativity and thinking of an image that the organic molecule might remind you of. For example, I always thought P-xylene reminded me of a bird.

The common names that I have seen constantly come up on exams are the following:

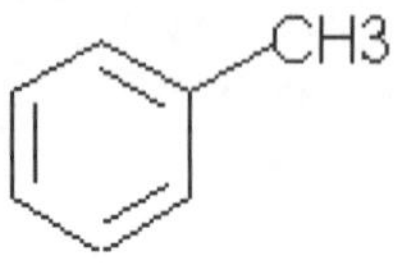

CH3

Toluene

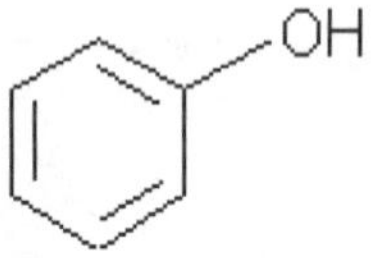

OH

Phenol

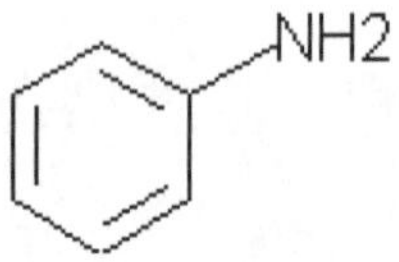

NH2

Aniline

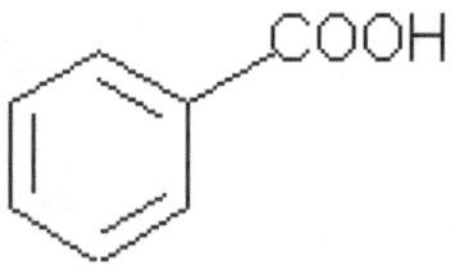

COOH

Benzoic Acid

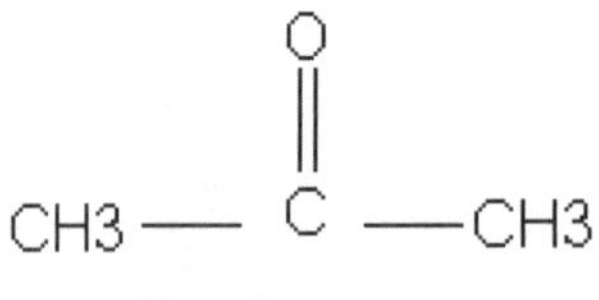

O

CH3 — C — CH3

Acetone

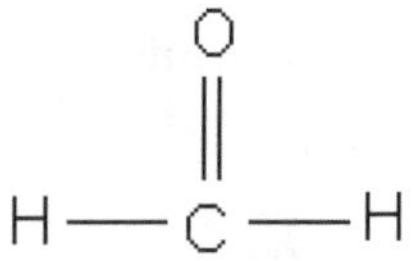

O

H — C — H

Formaldehyde

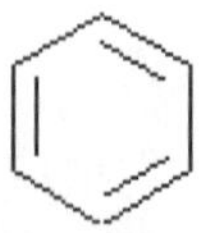

Benzene

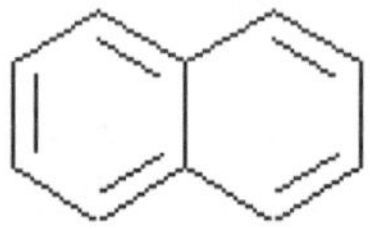

Naphthalene

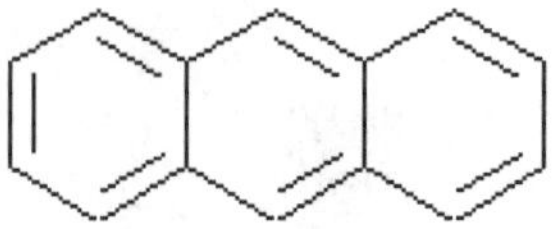

Antracene

3. FUNCTIONAL GROUPS

Functional groups are a part of organic chemistry that will haunt you for the entire duration of your chemistry education. For this reason it is a good idea to commit them to memory sooner rather than later. The best way to digest functional groups is to break them up into two categories. The way that I break them up is by those of which that do not have a C=O and those that do have a C=O. By doing this you will be able to quickly recognize functional groups based on whether or not a C=O is present.

This section is organized first by those without C=O and then those with C=O. Within each part, we'll talk about how to recognize and name each functional. Please note that an 'R' in is used as a variable to represent the 'Rest' of the molecule in question.

3a. Functional groups without C=O

Alcohol (R-OH)

Recognize: Whenever you see is an –OH without the presence of a C=O

Useful hint: A primary alcohol is when the OH is connected to a carbon that is connected to one other carbon. A secondary alcohol is when the OH is connected to a carbon that is connected to two other carbons etc.

Name: Just like how the world Alcohol ends in –ol, so does the name of alcohols. To name an alcohol use the formula: prefix + an + ol ending. The prefix is the number of carbons in the molecule and the ending for an alcohol is –ol.

Example:

C—C—C—C—OH **Butanol**

Ether (R-O-R')

Recognize: Whenever there is an O in the middle of a carbon chain without the presence of a C=O

Useful hint: Think of Easter (ether) bunnies —O—

Name: List the substituents on each side of the O in alphabetical order and then add the word ether

Example:

C—O—C—C **Ethylmethyl Ether**

Peroxide (R-O-O-R')

Recognize: Whenever there is an O-O in the middle of a carbon chain without the presence of a C=O

Name: list the substituents on each side of the O-O in alphabetical order and then end with the word peroxide

Example:

C—O—O—C—C **Ethylmethyl Peroxide**

Amine (N-R)

Recognize: Whenever there is nitrogen without the presence of a C=O

**Note that a primary amine is connected to one other carbon, secondary connected to two, tertiary connected to 3

Name: List the substituents that are around the nitrogen in alphabetical order and then end the name with amine.

```
                                  You
                                   |
Useful Hint: The name of    DO — N — KnowWhat   is
DoYouKnowWhatAmine          (Do you know what I mean)
```

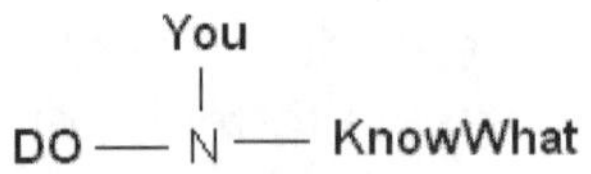

Example:

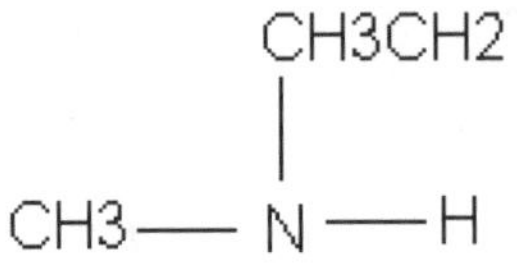

```
           CH3CH2
             |
CH3 —— N —— H      Ethylmethylamine
```

3b. With C=O

```
              O
              ||
Aldehyde  (C -H)
```

Recognize: Whenever there is a C=O on the end of an organic chain.

Useful hint: Look for a "CHO" at the end of the molecule

Name: The word aldehyde starts with al-, and so does the ending of aldehyde names. So use the formula prefix + an + al ending.

Example:

```
        O
        ‖
H—C—C—C
```
Propanal

```
          O
          ‖
```
Ketone (C-C-C)

Recognize: Whenever there is a C=O in the middle of an organic chain with carbons on both sides

Name: Just as the word ketone ends with –one, so does the ending of ketone names. So use the formula prefix + an + -one ending.

Example:

```
            O
            ‖
C—C—C—C—C
```
Pentanone

```
         O
         ‖
```
Carboxylic acid (C -OH)

Recognize: Whenever there is an OH attached to a C=O

Useful Hint: Note the difference between acids and alcohols. They both have the OH but alcohol does not have the C=O and acids do

Name: Organic acids end in –oic and are followed by the word acid. Use the formula prefix + an + oic ending + acid.

Example:

```
            O
            ‖
C—C—C—C—OH
```
Butanoic Acid

Ester (R-C(=O)-OR)

Recognize: Whenever there is an O attached to a C=O

Useful Hint: Look for a "COOCH" at the end of the molecule

Naming: The O in the chain disrupts the longest chain and causes the shorter half to be a substituent and the longer half to be the longest chain. First you must name the shorter end as the substituent then name the longer end. The longer end is named using the formula prefix + an + ending with the ending being –oate.

Example:

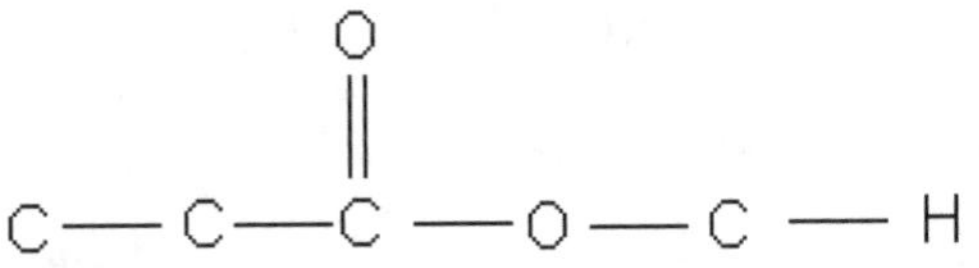

C — C — C(=O) — O — C — H **Methyl Propanoate**

Amide (C(=O)-N)

Recognize: Whenever there is an N attached to a C=O

Name: The name of an amide ends with –amide. Use the formula prefix + an + ending

Example:

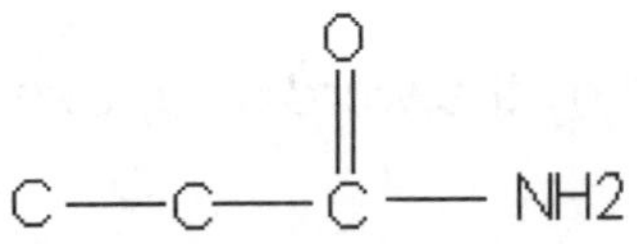

C — C — C(=O) — NH2 **Propanamide**

Dioic acid (OH-C(=O)-c_n-C(=O)-OH)

Recognize: Whenever there are two acid groups in the same molecule

Useful hint: Pneumonic Device: **O**h **M**y **S**uch **G**ood **A**pple **P**ie

Name: The name is based on how many carbons there are in between the two acid groups. Use the following names which correspond to the specific number of carbons in the middle starting at 0. The ending of the name is acid.

0 carbons – oxalic
1 carbon – malonic
2 carbons – succinic
3 carbons – glutaric
4 carbons – adipic
5 carbons – pimelic

Example:

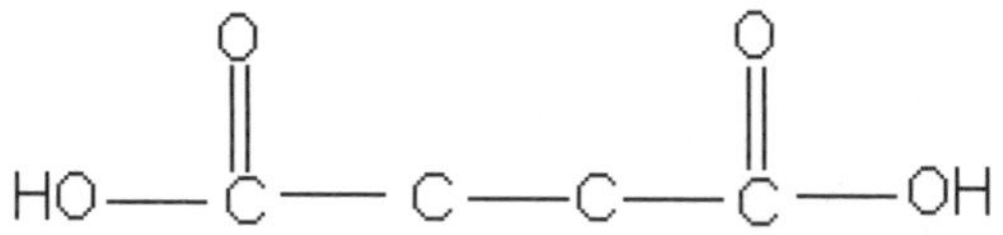

HO—C(=O)—C—C—C(=O)—OH **Succinic Acid**

4. CYCLIC (AROMATIC) MOLECULES

4a. Cyclic Nomenclature

For your intents and purposes in introductory chemistry, the words cyclic and aromatic are synonymous. When naming a molecule with a cyclic ring, the ring is always considered the longest chain. You never want to break a ring in half or divide it in any way. Rings are named the same way as regular straight chains, but the only difference is that you need to put the word cyclo in front of the name. When naming the substituents around the ring, the same rules apply. You must number the substituent, designate how many carbons are present, and end the name with –yl.

When numbering the ring, always start at a carbon that has a substituent already on it. This way you will be able to have the smallest numbered substituents possible. Then number in the direction that will leave you with the smallest numbers for the other substituents. Another way of designating where the substituents are is by using special prefixes. When two substituents are one carbon away on the ring you use the prefix ortho. When two substituents are two carbons away you use the prefix meta. When two substituents are three carbons away you use the prefix para.

Refer to the following diagram for a summary:

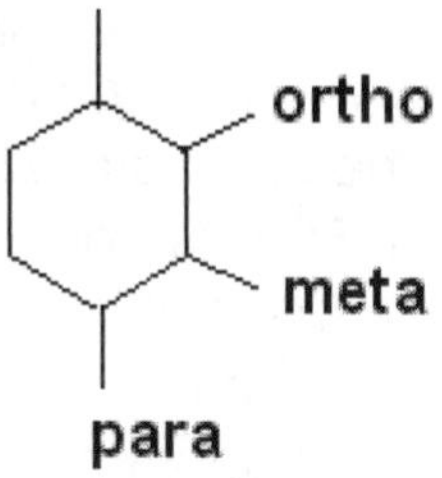

ortho
meta
para

Examples of naming cyclic molecules:

CH3

CH2CH2CH3

1-methyl-2-propylcyclopropane

CH2CH3

CH3

1-ethyl-2-methylcyclohexane

4b. Difference between benzene and Cyclohexane

An important distinction that must be made clear is the difference between benzene and a cyclohexane molecule; both of which are written with hexagons and so are easily confused. A benzene molecule has the formula C_6H_6 and has alternating double and single bonds. A cyclohexane molecule has the formula C_6H_{12} and has all single bonds. A lot of people know that benzene has alternating double bonds, but get confused when it is written with a circle in the middle. This circle is simply shorthand for writing in the double bonds. Be sure you recognize the difference between:

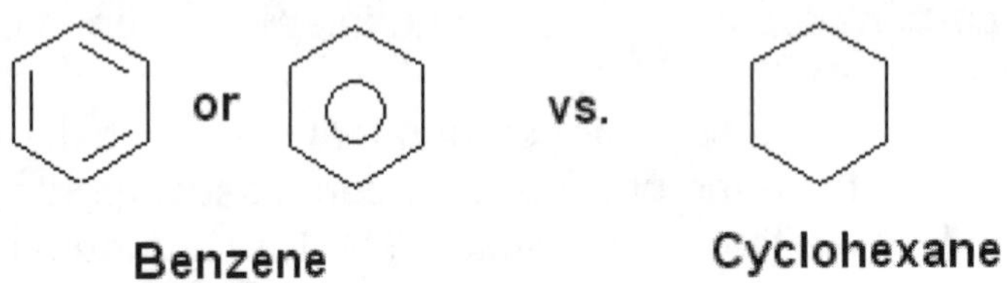

Benzene **Cyclohexane**

5. POLYMERS

5a. Homopolymers

A polymer is a repeating unit of monomers. The typical questions that are asked about polymers are in the form of identification. As long as you can recognize the different polymers that were talked about in class you will be fine. The way that I would go about being able to recognize them is by the one characteristic that sets each one apart from one another. For example, polyvinylchloride is the only polymer that has chlorine in it so anytime you see chlorine you should assume that it has to be polyvinylchloride. The same rationale holds true for a polymer like Teflon. Teflon is the only polymer that has fluorine in it and therefore anytime you see fluorine you can recognize it as Teflon.

5b. Condensation Copolymers

When there is a C=O in the polymer it is known as a condensation copolymer. The way that I would handle these types of polymers is to first locate and circle the C=O. You then want to look to see what is directly attached to the C=O. Depending on what is attached you will be able to determine the type of condensation copolymer. The following is a list of the possibilities of what can be attached to the C=O along with the condensation copolymer that it refers to.

O – Polyester
N – Polyamide
O and N - Ureathane

6. ISOMERS

Questions involving isomers are usually asked in the form of a multiple choice question. The word isomer refers to any two structures that has the same chemical formula but are oriented differently. A good way to remember this is that the Latin root "iso" means "the same." There are three different types of isomers depending on the way that a second molecule is oriented differently compared to the original molecule.

6a. Constitutional Isomers

A constitutional isomer is an isomer that has the same chemical formula but has different bonding relationships between the atoms. In other words, it still has the same amount of each element, but the elements are arranged and bonded differently. A perfect example of constitutional isomers is a primary alcohol versus a secondary alcohol. As you can see in the example below, each has the formula C_3H_7OH but the OH is attached to the third carbon in the first molecule and attached to the second carbon in the second molecule. Therefore they have the same chemical formula but have different bonding relationships making them constitutional isomers:

```
                                    OH
                                    |
C—C —C—OH          Vs.        C—C—C
```

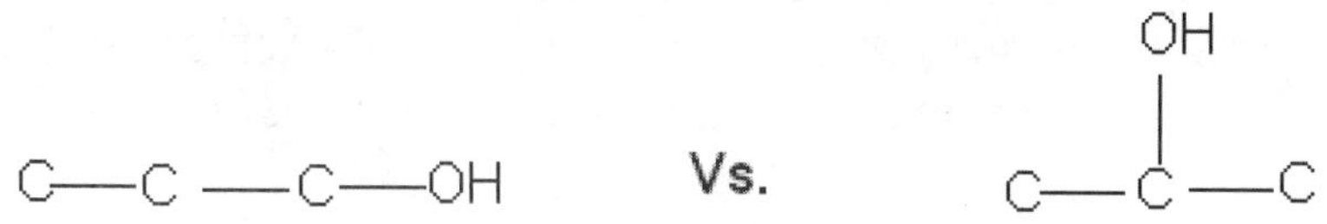

6b. Geometric Isomers

A geometric isomer is an isomer that has the same structural formula but has a different orientation at a double bond. Unlike single bonds, double bonds cannot rotate freely and are locked into place. This means the groups on either side of a double bond are fixed in place relative to each other. Geometric isomers are quite simply two of the same molecules but with different orientations about the double bond. A perfect example of geometric isomers would be cis and trans double bonds. Cis, as I like to think of as sisters, is when the longest chain continues on the same side of the double bond. Another way to look at it is that the longest chain forms a C shape just like how the word cis starts with a C. Trans, which is a latin root for 'across' is when the longest chain continues on opposite sides of the double bond. Another way to look at it is that the longest chain forms the letter S. An example of cis vs. trans is illustrated below:

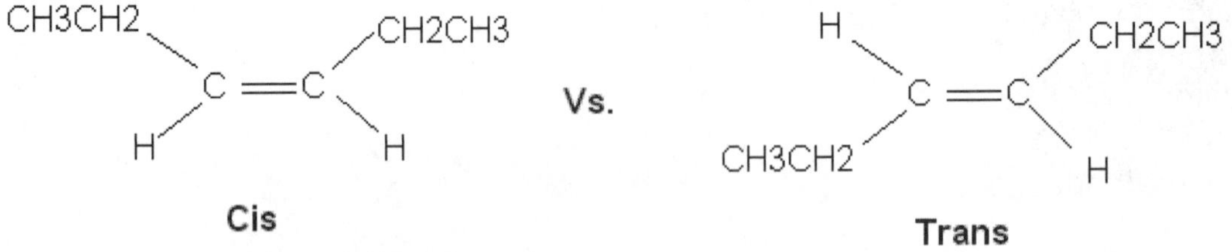

CH3CH2 CH2CH3
C=C
H H

Cis

Vs.

H CH2CH3
C=C
CH3CH2 H

Trans

**Hint: When a question is referring to isomers and there is a double bond in the molecule, it will most likely be talking about geometric isomers. In other words, a double bond should serve as a red flag for recognizing geometric isomers.

6c. Stereoisomers

Stereoisomers are two molecules with the same chemical formula but a different 3D orientation of groups around a central carbon. In order to have stereoisomers, a stereocenter must be present in a molecule. A stereocenter is a central carbon with the criterion that there must be four DIFFERENT groups attached to it, as illustrated below:

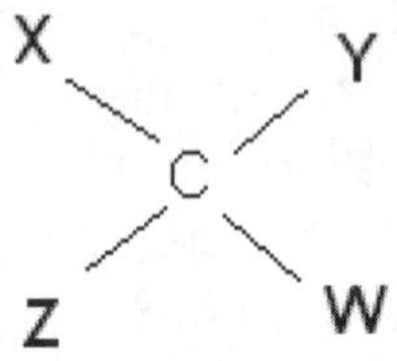

X Y
C
Z W

The orientation of the four different things attached to the central carbon will determine the stereoisomer. If two stereoisomers are actually mirror images of each other, they form a pair of enantiomers. The example below illustrates an example of enantiomers:

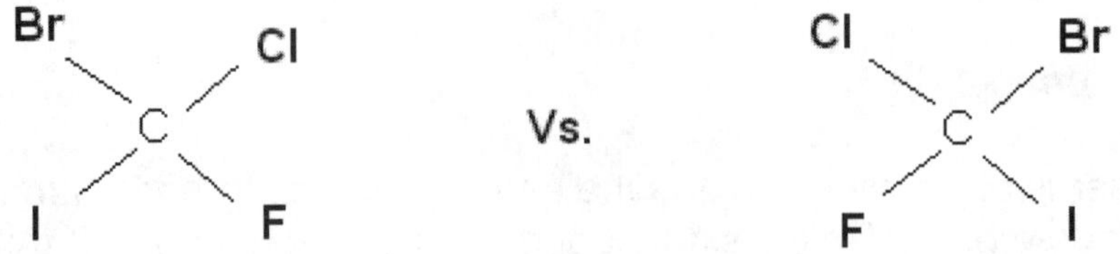

Br Cl
C
I F

Vs.

Cl Br
C
F I

Note how one molecule looks like the other reflected in a vertical mirror, so these stereoisomers are in fact enantiomers.

7. ORGANIC REACTIONS

7a. Addition Reactions

An addition reaction occurs anytime the starting reactant has a double or triple bond. When a double or triple bond is present, the molecule is said to be unsaturated, which means we could still add more elements such as hydrogen to it. Therefore, recognize anytime the starting material is unsaturated that it is most likely going to undergo an addition reaction.

With an addition reaction, the double bond breaks which allows two more atoms to be added onto the molecule. Typically what happens is an alkene reacts with something like HCl, causing the double bond to break and the H and the Cl to be added to the molecule on the two carbons that once had the double bond. The reason for this is that once the double bond breaks, each carbon involved no longer has 4 bonds and it needs to compensate for that. One important thing to note is that the Cl goes to the carbon with the least amount of hydrogens and the hydrogen goes to the carbon with the most amount of hydrogens. A good way to think of that is "Hydrogen goes where hydrogen is" or "Those that have will get, and those that don't have will not".

An example of an addition reaction:

```
        H    H    H                                      H    H    H
        |    |    |                                      |    |    |
   H —— C —— C == C —— H   +   HCl   ———————>   H —— C —— C —— C —— H
        |                                                |    |    |
        H                                                H    Cl   H
```

7b. Elimination Reactions

Elimination reactions are the opposite of addition reactions. In an elimination reaction, instead of starting with a double bonded molecule, you end with a double bonded molecule. Two elements get eliminated from the starting reactant and in their place, in order to compensate for the loss of the stable octet, a double bond is formed. These kinds of reactions are not spontaneous, meaning they do not happen on their own. Therefore there needs to be some kind of catalyst that will cause this kind of reaction to happen. You will see this in the reaction written as cat for catalyst and Δ for heat.

An example of an elimination reaction:

$$\begin{array}{ccccccc} & H & & H & & H & \\ & | & & | & & | & \\ H- & C & - & C & - & C & -H \\ & | & & | & & | & \\ & H & & Cl & & H & \end{array} \xrightarrow{\text{cat}, \Delta} \begin{array}{ccccccc} & H & & H & & H & \\ & | & & | & & | & \\ H- & C & - & C & = & C & -H \\ & | & & & & & \\ & H & & & & & \end{array} + HCl$$

7c. Substitution Reactions

Substitution reactions are a little different than addition or elimination reactions. In substitution reactions, nothing is added or eliminated but rather something is replaced. Therefore there is still the same amount of elements in the beginning and the end of the reaction. The bonds remain unchanged and usually reactants undergoing substitution reactions start as a saturated compound and remain as a saturated compound:

An example of a substitution reaction:

$$\begin{array}{ccccccc} & H & & H & & H & \\ & | & & | & & | & \\ H- & C & - & C & - & C & -Cl \\ & | & & | & & | & \\ & H & & H & & H & \end{array} \xrightarrow{OH-} \begin{array}{ccccccc} & H & & H & & H & \\ & | & & | & & | & \\ H- & C & - & C & - & C & -OH \\ & | & & | & & | & \\ & H & & H & & H & \end{array} + HCl$$

7d. Redox Reactions

Oxidation

In an oxidation reaction, carbon atoms form an additional bond to oxygen atoms resulting in a C=O molecule. The way to recognize these reactions is when there are reagents like H_2SO_4 and K_2CrO_7 used. Also, sometimes the reaction is denoted using the letters 'ox' above the arrow in the reaction.

With primary alcohols, the sequence of oxidation always goes from alcohol into an aldehyde which then gets oxidized into an acid. If it is a secondary alcohol it becomes a ketone but cannot then be oxidized further into a carboxylic acid.

Reduction

In a reduction reaction, carbon atoms form fewer bonds to oxygen atoms, resulting in loss of a C=O bond. The way to recognize these reactions is when there are reagents like Pd or when the letters 'red' appear above the arrow of the reaction.

The sequence of reduction is the opposite of oxidation. Acids get reduced to Aldehydes which get reduced to primary alcohols or secondary alcohols depending on where the oxygen atom is.

7e. Ester Synthesis

One last organic reaction that is usually always seen on exams is ester synthesis. In order to understand ester synthesis you must first recall what an ester is. An ester is a functional group with an oxygen atom attached directly to a C=O. In order to make an ester you must combine an acid and an alcohol. What happens is that the hydrogen on the end of the acid along with the OH on the alcohol combine to form a water molecule, causing the two compounds to come together to form an ester. My best advice is to circle the H and the OH in acid and the alcohol respectively and mentally block them out of your head. Visualize the remainder of both of the reactants coming together and that is your final answer. This is illustrated in the following example:

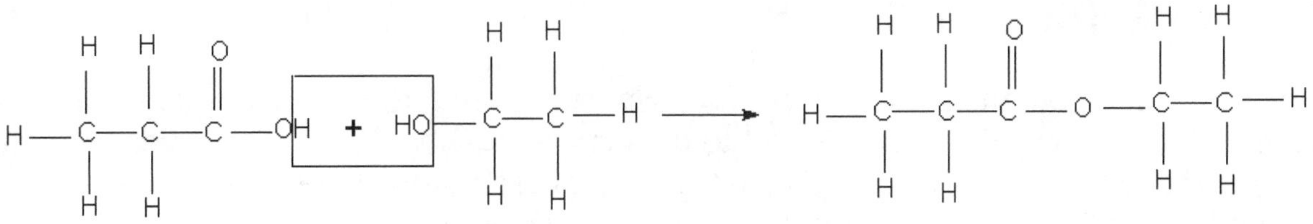

SUMMARY

Some people will argue that organic chemistry is the hardest topic that you will ever face in introductory chemistry. (In fact, it is often offered as a separate course to be taken AFTER Intro chemistry). Other people will argue that it is the easiest. It basically comes down to your own personal preference. A lot of people like organic chemistry because there is no math involved. You might have noticed that there was absolutely no math involved in this entire chapter.

As you have seen from this chapter, the most important way of approaching organic chemistry is to be able to recognize the material. You must be able to recognize the type of question they are asking you and how to go about answering it. For nomenclature, take the very simple systematic approach to naming organic molecules and you should not have a problem with even the most complicated looking molecules. Be familiar with the common names enough to be able to recognize them and you should not have a problem with that either. For the functional groups, we talked about the different distinctions between all of them and how to recognize them. First ask yourself whether or not there is a C=O. If there is, ask yourself where is the C=O and what is attached to it. Be able to understand and differentiate between the three types of isomers. Also, have a general understanding of the different types of organic reactions (which you usually don't see until the second half of introductory chemistry). The bottom line is do not let organic chemistry scare you.

CHAPTER 13

THE PERIODIC TABLE

OVERVIEW

The periodic table is usually the first thing that comes to mind when thinking about the subject of chemistry. We have all heard about the periodic table from as far back as elementary school but probably never taken the time to really examine it. The periodic table tells us a lot of important pieces of information that are used throughout the course of chemistry. Some of the information is obvious, and other pieces are quite ambiguous. This chapter will walk you through exactly what you are expected to know and will also provide sample questions involving the table and the information that it contains.

1. VIEWING THE PERIODIC TABLE AS INDIVIDUAL COMPONENTS

1a: Atoms and Chemical Elements

Before talking about the periodic table as a whole, it is important to understand what the periodic table is composed of. In order to do this we must begin our discussion with the atom and subatomic particles. An atom is the smallest indivisible particle of matter. Atoms contain three types of subatomic particles known as protons, neutrons, and electrons. Protons are positively charged, neutrons are neutral, and electrons are negatively charged. In terms of location, protons and neutrons are located in the nucleus and electrons are located on the outside of the nucleus. A chemical element is a type of atom that contains a specific amount of protons in the nucleus. Chemical elements are distinguished from one another by the number of protons in the nucleus of each atom. Depending on the number of each subatomic particle, and the ratio of subatomic particles to each other, specific characteristics about a given atom can be determined.

1b: Atomic Number (Protons)

The amount of protons that an element contains is the atomic number of that element. Usually in the table, the number above each element is the atomic number of that element and so that tells us exactly how many protons are located in its nucleus, as shown here for carbon:

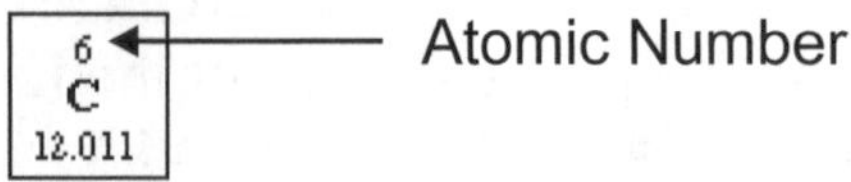

Looking closely at the periodic table, you will notice that each element is arranged in order of increasing atomic number, starting with hydrogen with an atomic number of 1.

1c: Atomic Mass (figuring out the number of neutrons)

The mass number corresponds to the number of protons plus the number of neutrons. When given the mass number, the number of neutrons can be easily calculated because of the relationship protons + neutrons = mass number. By subtracting both sides of the equation by the number of protons we obtain a new equation which states neutrons = mass number – protons.

However, it is crucially important to clear up a common source of confusion here. The mass number for a given atom is simply the number of protons plus number of neutrons, as stated above. But this is NOT what appears in the periodic table, and students often get these confused. And the reason for this is the existence of isotopes. There is a difference between the meanings of atomic mass and mass number. Atomic mass is the average weight of an element and the mass number is the total number of nucleons in the atom's nucleus.

The truth is that elements exist in several different forms, or isotopes. These are differing atoms of the same element, meaning they have the same number of protons, but which differ in the number of neutrons. Due to the fact that the sum of protons and neutrons is equal to the mass number, the atomic mass of the isotopes for a given element will be different. An example of isotopes is illustrated in the following example:

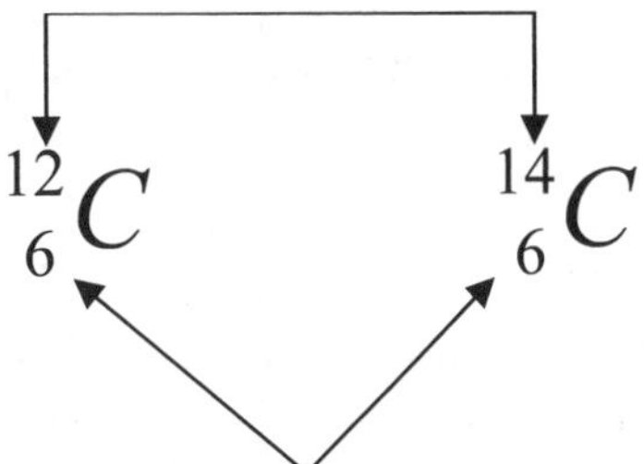

It is very important to note that the atomic mass that is written on the periodic table is really an average of all of the isotope masses that exist for that particular element, based upon the relative abundance of each isotope. This makes sense because if one isotope exists a lot more than another isotope, than the isotope that exists more should have more of an impact on the mass of that element.

The way to account for this is to use a formula as shown:

Mass of element = (Mass of Isotope 1)(relative abundance)
+ (Mass of Isotope 2)(relative abundance)

Example: Calculate the mass of Copper, given its two isotopes ${}^{63}_{29}Cu$ (mass = 62.93) and ${}^{65}_{29}Cu$ (mass = 64.93) have relative abundances 69.2 % and 30.8 % respectively.

Step 1: Write the equation:

Mass of element = (Mass of Isotope 1)(relative abundance)
+ (Mass of Isotope 2)(relative abundance)

Step 2: Convert the % abundances to relative abundances by dividing by 100:

69.2 % = 0.692 and 30.8 = 0.308

Step 3: Plug in given values:

Mass of copper = (62.93)(0.692) + (64.93)(0.308)

Step 4: Solve for unknown:

Mass of copper = 63.55

Looking on the periodic table you will notice the atomic mass for copper is given as 63.55, which is equal to the value calculated. This demonstrates how the atomic mass of each element is determined based on the isotopes and their respective abundances.

Instructors typically ask a few questions like this on their tests. I must warn you, not every question involves calculating the atomic mass of the element given the different isotopes. Sometimes instructors give you the atomic mass of the element along with one isotope and ask you to solve for the other isotope. The best way to go about doing this is to assign an "x" to whichever quantity the question is asking for and solving for that variable using simple algebra. Just remember, every variable in a formula is how many different ways you can be asked question on a topic!

1D: Ions and Charges (electrons)

All the elements on the periodic table are neutral. This means that an element does not have a net charge associated with it. In order for this to be the case, it must have an equal number of protons and electrons. Since protons are positively charged and electrons are negatively charged, equal amounts of each cancel each other out and there is no net charge.

However, what happens when there is not an equal amount of protons and electrons? The number of protons never changes and is always the same for a specific element. The number of electrons can in fact change and cause something called an **ion** to form. An ion is a charged element that does not have an equal amount of protons and electrons.

Positively Charged Ions

A positively charged ion is one in which there is more protons than electrons. Since there are more positive charges than negative charges, a net positive charge results. We call positively charged ions **cations**. A good way to remember this term is:

A dead cat is not a good (positive) thing

Another way is to think of the 't' in cation as looking a bit like a positive charge.

Negatively Charged Ions

A negatively charged ion is one in which there is more electrons than protons. Since there are more negative charges than positive charges, a net negative charge results. We call negatively charged ions **anions.** A good way to remember this term is to think of '<u>a</u> <u>n</u>egative <u>ion</u>'.

The question is how do we determine which elements become cations and which become anions? The answer is very simple and is easily explained using the periodic table. We will expand on this in the section below.

2. VIEWING THE PERIODIC TABLE AS A WHOLE

2A: The Soap Opera of chemistry

Think of chemistry as a soap opera. The ultimate goal of every element is obtain 8 valence electrons. The noble gases - the elements on the far right side of the periodic table in column 18 - already all have a stable octet of electrons; so they are completely content and have no need to bother with any other. However every single other element in the periodic table is not as fortunate and so is jealous. In other words, they will do anything in their power to become just like a noble gas.

The only way for these elements to become like a noble gas on their own is by gaining or losing electrons, depending on whichever would give the element the fastest route to obtaining the noble gas character. The particular element's position in the periodic table will determine which route would be best and how many electrons the element will need to gain or lose.

Elements that need to gain electrons

The elements to the immediate left of the noble gasses need to gain just one electron to become a noble gas. Group 17, also known as the halogens, is a group of elements that are immediately next to the noble gases as seen in the example below:

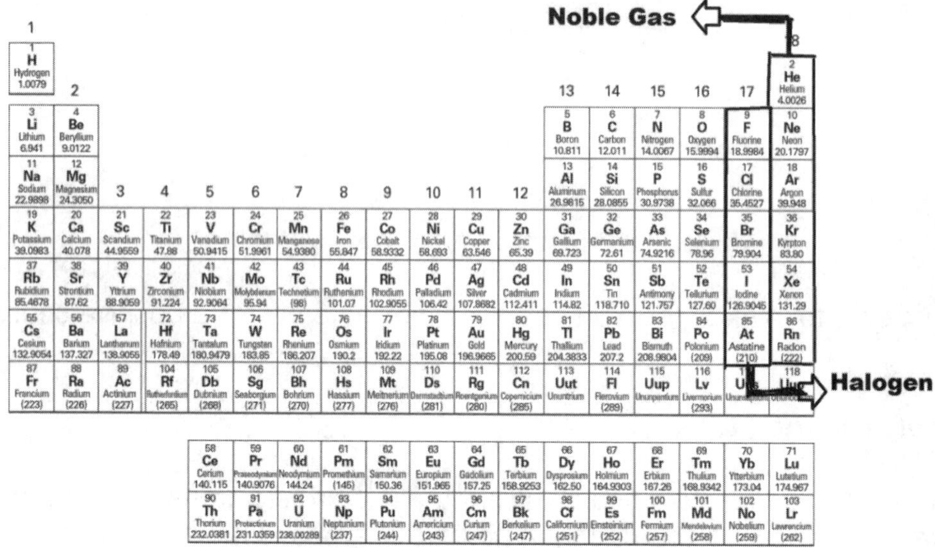

Since the halogens are only one step away from becoming a noble gas, they only need to gain one electron. By doing this, each halogen cn obtain the same amount of electrons as a noble gas. Remember that electrons are negatively charged, so by gaining one electron all of the halogens will pick up a -1 charge. For this reason the halogens become ions such as F^{-1},Cl^{-1}, Br^{-1}, I^{-1} etc.

The next group over is the group with oxygen at the top. This group is number 16 and each member contains 6 valence electrons. On the periodic table, this group is two steps away and from becoming a noble gas and so must gain 2 electrons to do so. This results in a -2 charge; therefore the ions of group 16 consist of O^{-2} S^{-2} Se^{-2} etc.

The next group over is the group with nitrogen at the top. This group is number 15 and each member contains 5 valence electrons. On the periodic table, this group is three steps away and from becoming a noble gas and so must gain 3 electrons to do so. This results in a -3 charge; therefore the ions of group 15 consist of N^{-3} P^{-3} As^{-3} etc.

Elements that need to lose electrons

As we keep moving further and further away from the noble gases, starting with group 14, there becomes a point where it no longer makes the most sense to gain electrons. Group 14 is unique because it is equally distant from the noble gases whether it loses or gains electrons. For this reason, this group can technically gain 4 electrons or lose 4 electrons making each one of the ions in the group +/- 4. However, group 14 elements do not usually exist as ions In terms of ions.

Continuing across, group 13 elements have 3 valence electrons. It would not make any sense for this group to gain 5 electrons when it could much easier just lose 3 electrons. Since group 13 elements are three groups away from the noble gases in the other direction they prefer to lose 3 electrons instead. Keep in mind that these elements are now losing electrons and so are becoming positively charged as opposed to negative. In other words, the ions of group 13 consist of B^{+3} Al^{+3} Ga^{+3} In^{+3} etc.

Skipping past groups 12 to 3 which are the transition metals which we will be talking about next, we now arrive at group 2. Group 2, which is also known as the alkaline earth metals, contains elements that contain 2 valence electrons and is located 2 steps away from the noble gases in the reverse direction. For this reason elements in this group loses 2 electrons and so obtain a +2 charge. The ions are Be^{+2} Mg^{+2} Ca^{+2} Sr^{+2} Ba^{+2} etc.

Last we arrive at group 1. Group 1 elements are referred to as the alkali metals. These elements contain 1 valence electron and are 1 group away from the noble gasses in the reverse direction. Therefore these elements lose 1 electron and obtain a +1 charge. The ions are Li^{+1} Na^{+1} K^{+1} Rb^{+1} Cs^{+1} etc.

SUMMARY OF THE PERIODIC TABLE SOAP OPERA

Elements on the periodic table want to become a noble gas to gain a stable octet of 8 valence electrons. How close the group is to the noble gases will determine how many electrons will be lost or gained. The following periodic table summarizes the charges of ions from the groups that we previously just discussed

+1	+2											+ 3	± 4	- 3	- 2	- 1	
1	2	3	4	5	6	7	8	9	10	11	12	13	14	15	16	17	18
1 H Hydrogen 1.0079																	2 He Helium 4.0026
3 Li Lithium 6.941	4 Be Beryllium 9.0122											5 B Boron 10.811	6 C Carbon 12.011	7 N Nitrogen 14.0067	8 O Oxygen 15.9994	9 F Fluorine 18.9984	10 Ne Neon 20.1797
11 Na Sodium 22.9898	12 Mg Magnesium 24.3050											13 Al Aluminum 26.9815	14 Si Silicon 28.0855	15 P Phosphorus 30.9738	16 S Sulfur 32.066	17 Cl Chlorine 35.4527	18 Ar Argon 39.948
19 K Potassium 39.0983	20 Ca Calcium 40.078	21 Sc Scandium 44.9559	22 Ti Titanium 47.88	23 V Vanadium 50.9415	24 Cr Chromium 51.9961	25 Mn Manganese 54.9380	26 Fe Iron 55.847	27 Co Cobalt 58.9332	28 Ni Nickel 58.693	29 Cu Copper 63.546	30 Zn Zinc 65.39	31 Ga Gallium 69.723	32 Ge Germanium 72.61	33 As Arsenic 74.9216	34 Se Selenium 78.96	35 Br Bromine 79.904	36 Kr Kyrpton 83.80
37 Rb Rubidium 85.4678	38 Sr Strontium 87.62	39 Y Yttrium 88.9059	40 Zr Zirconium 91.224	41 Nb Niobium 92.9064	42 Mo Molybdenum 95.94	43 Tc Technetium (98)	44 Ru Ruthenium 101.07	45 Rh Rhodium 102.9055	46 Pd Palladium 106.42	47 Ag Silver 107.8682	48 Cd Cadmium 112.411	49 In Indium 114.82	50 Sn Tin 118.710	51 Sb Antimony 121.757	52 Te Tellurium 127.60	53 I Iodine 126.9045	54 Xe Xenon 131.29
55 Cs Cesium 132.9054	56 Ba Barium 137.327	57 La Lanthanum 138.9055	72 Hf Hafnium 178.49	73 Ta Tantalum 180.9479	74 W Tungsten 183.85	75 Re Rhenium 186.207	76 Os Osmium 190.2	77 Ir Iridium 192.22	78 Pt Platinum 195.08	79 Au Gold 196.9665	80 Hg Mercury 200.59	81 Tl Thallium 204.3833	82 Pb Lead 207.2	83 Bi Bismuth 208.9804	84 Po Polonium (209)	85 At Astatine (210)	86 Rn Radon (222)
87 Fr Francium (223)	88 Ra Radium (226)	89 Ac Actinium (227)	104 Rf Rutherfordium (265)	105 Db Dubnium (268)	106 Sg Seaborgium (271)	107 Bh Bohrium (270)	108 Hs Hassium (277)	109 Mt Meitnerium (276)	110 Ds Darmstadtium (281)	111 Rg Roentgenium (280)	112 Cn Copernicium (285)	113 Uut Ununtrium	114 Fl Flerovium (289)	115 Uup Ununpentium	116 Lv Livermorium (293)	117 Uus Ununseptium	118 Uuo Ununoctium

58 Ce Cerium 140.115	59 Pr Praseodymium 140.9076	60 Nd Neodymium 144.24	61 Pm Promethium (145)	62 Sm Samarium 150.36	63 Eu Europium 151.965	64 Gd Gadolium 157.25	65 Tb Terbium 158.9253	66 Dy Dysprosium 162.50	67 Ho Holmium 164.9303	68 Er Erbium 167.26	69 Tm Thulium 168.9342	70 Yb Ytterbium 173.04	71 Lu Lutetium 174.967
90 Th Thorium 232.0381	91 Pa Protactinium 231.0359	92 U Uranium 238.00289	93 Np Neptunium (237)	94 Pu Plutonium (244)	95 Am Americium (243)	96 Cm Curium (247)	97 Bk Berkelium (247)	98 Cf Californium (251)	99 Es Einsteinium (252)	100 Fm Fermium (257)	101 Md Mendelevium (258)	102 No Nobelium (259)	103 Lr Lawrencium (262)

2b: The transition metals

You are probably wondering why we skipped over the group of elements in the middle of the periodic table. These elements are known as the transition metals and are located in groups 3 to 12. The transition metals do not follow the general theme of gaining or losing electrons depending on their location relative to the noble gases. The transition metals, unlike the other elements, actually have more than one charge associated with them. You will not be responsible for knowing every transition metal on the periodic table, just the ones in the top row. Below are the transition metals that you will be expected to know along with their respective charges:

Element	Symbol	Charge 1	Charge 2
Chromium	Cr	2+	3+
Manganese	Mn	2+	3+
Iron	Fe	2+	3+
Cobalt	Co	2+	3+
Nickel	Ni	2+	4+
Copper	Cu	1+	2+
Mercury	Hg	1+	2+
Gold	Au	1+	3+
Cerium	Ce	3+	4+

2c: Metals vs. Non-metals

Now that we have discussed the charges associated with the different elements, it is also important to understand the difference between metals and non-metals. In some periodic tables you will notice that a staircase is drawn in on the right potion of the table. This staircase is the dividing line between the metals and the non-metals. Metals are the elements located to the left side of the staircase and non-metals are the elements located to the right. Metals lose electrons and become positively charged. Non-metals gain electrons and become negatively charged. The elements located directly on the staircase are known as metalloids and experience characteristics of both metals and non-metals. The difference between metals and nonmetals will be particularly important in the chapter where we talk about bonding and chemical nomenclature. For now, just understand that there is a difference between the two and how to identify them based on their position relative to the staircase. Below is a summary of metals and non-metals.

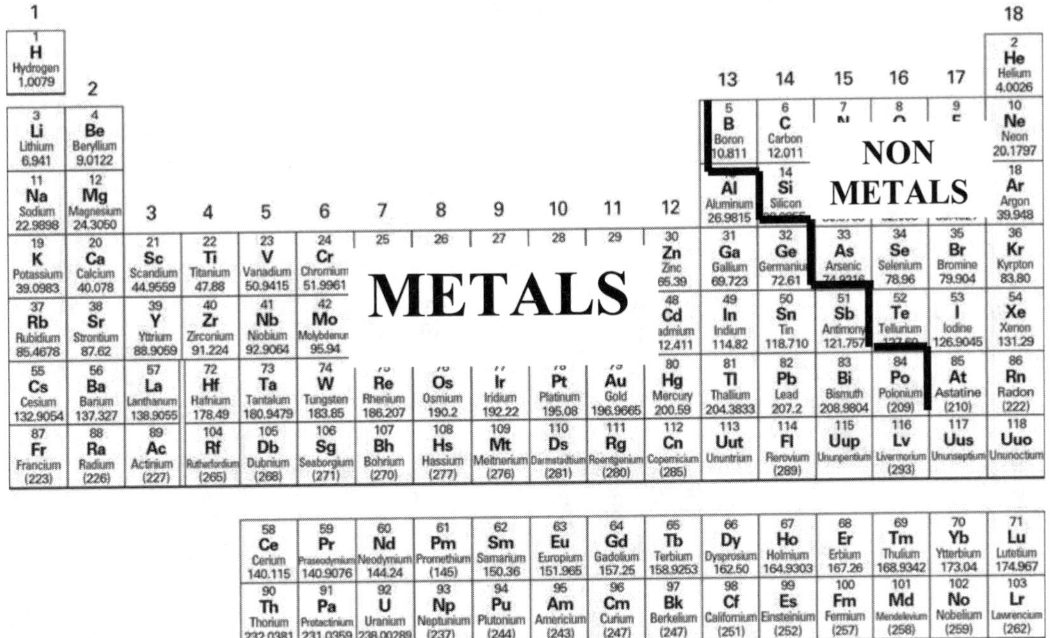

2d: Trends

There are important trends that you will need to know about the periodic table. The three trends are electronegativity, ionization energy, and atomic radius. The three trends are discussed in the remainder of this section.

Electronegativity

Electronegativity is a fancy word with a very simple definition. Electronegativity refers to an element's ability to attract an electron, when in a bond with another element. Some elements have a much higher need for electrons than others. As you move to the right of the periodic table electronegativity increases. As you move upwards on the periodic table electronegativity increases. As a result there is a trend. In the uppermost right hand corner elements have the highest desire for electrons and therefore have the highest electronegativity. Fluorine in fact has the highest electronegativity of all the elements. Keep in mind that the noble gases are not included in this trend because as we know they have no need to gain or lose any electrons with their stable octet. The following periodic table summarizes electronegativity.

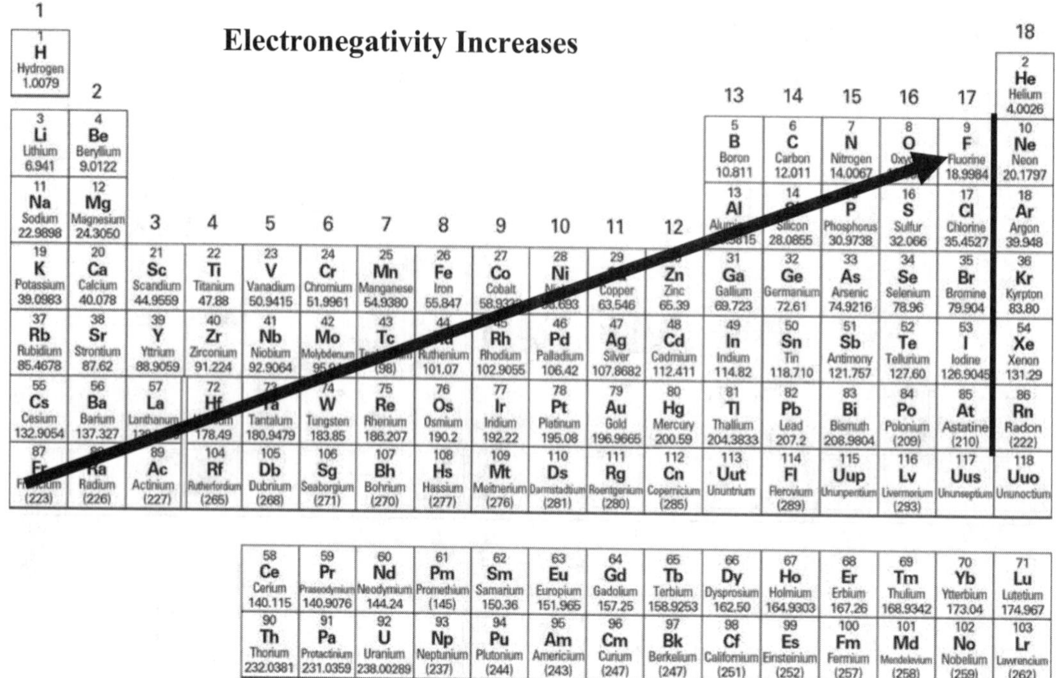

Ionization Energy

Ionization energy refers to the amount of energy required to remove an electron. The higher the ionization energy, the more difficult it is to remove an electron from an element. The trend for ionization energy is the same for that of electronegativity in that it also increases up and to the right. Meaning, Fluorine is the element on the periodic table with the highest ionization energy and is therefore the hardest element to remove an electron from. Again, keep in mind that the noble gases are not included in this trend. The following periodic table summarizes the trend for ionization energy.

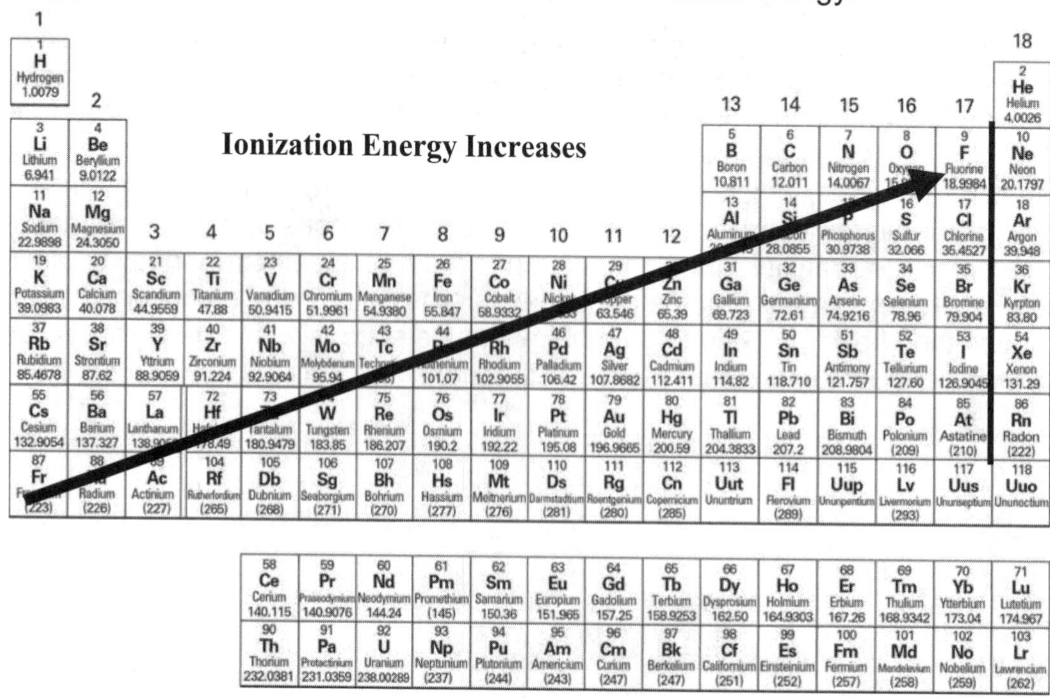

Atomic Radius

The last trend that you need to know is atomic radius. Atomic radius refers to the overall size of the element. As you go down a column, the radius of the element increases. As you go to the left across a row, the radius of the element increases. Therefore going down and to the left is where you will find the largest element. The largest element on the periodic table is Francium. The following periodic table summarizes the trend for atomic radius.

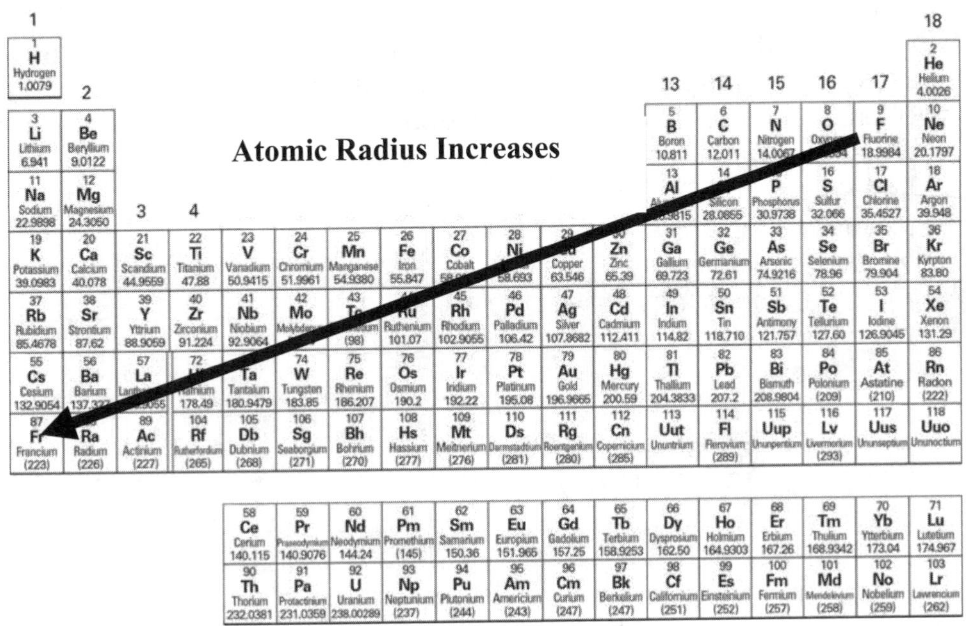

Sample Test Question:

In terms of an exam, the types of questions that are asked about the periodic table trends are usually in the form of multiple choice questions. An example of a question that an instructor would ask is the following:

Which of the following elements has the highest electronegativity?
A) Na B) Fe C) Al D) I e) Cl

The answer here would be chlorine. The best way to go about answering questions like the one above is to map out where each choice is on the periodic table. After putting a star or some sort of indication on the periodic table for each answer choice, you then want to remind yourself that electronegativity increases as you go up and to the right. Therefore the element that is the most up and to the right is Chlorine, choice E.

3. CALCULATIONS USING THE PERIODIC TABLE

A lot of calculations that we do involve using the molecular mass of a compound; in order to convert between grams and moles. A mole by definition is 6.02×10^{23} atoms. The mole is a very important unit and will be seen a lot in your study of chemistry.

3a: Calculating the molecular mass

A molecule is a term used for when more than one element is bonded together to form a compound. In order to calculate the molecular mass of any compound, you must add the atomic masses of all of the elements that are in the molecule. Remember, the atomic mass of each element is the number written below the element on the table.

Example: Calculate the molecular mass of NaCl

Step 1: Find the atomic mass of each element:
Na = 22.99
Cl = 35.45

Step 2: Add all of the masses together:
22.99 + 35.45 = 58.44 g

The only thing that you have to be careful of is when there is more than one of the same element present in a compound. When this happens, you must remember to multiply the mass of the element by how many times that element is present in the compound.

Example: Calculate the molecular mass of $BaCl_2$

Step 1: Find the atomic mass of each element
Ba = 137.3
Cl = 35.45 × 2 = 70.9 (since there are two chlorines)

Step 2: Add all of the masses together
137.3 + 70.9 = 208.2 g

3b: Converting between grams and moles

Now that we are comfortable calculating the molecular mass, we can use this to covert between grams and moles. Whenever you are trying to convert between grams and moles you should use dimensional analysis. The molecular mass is grams per mole of a compound which will serve as our conversion factor between the two units.

Example: How many moles are in 72g of H_2O?

Step 1: Always start with the number that you are given:

72g H_2O

Step 2: Use dimensional analysis to convert from grams to moles

$$72\text{ g H}_2\text{O}\left(\frac{\text{mol H}_2\text{O}}{\text{g H}_2\text{O}}\right)$$

Step 3: Plug in the calculated molecular mass of water and solve

$$72\text{ g H}_2\text{O}\left(\frac{\text{mol H}_2\text{O}}{\text{g H}_2\text{O}}\right) = 4\text{ mol H}_2\text{O}$$

3c: Converting between grams and number of atoms

Whenever you are asked to calculate the number of atoms present, you must first calculate the number of moles. You cannot go straight from grams to atoms. It is also important to keep in mind that there are 6.02×10^{23} atoms in one mole.

Example: How many atoms are present in 0.035 g of KCl?

Step 1: Always start with the number that you are given:

0.035 g KCl

Step 2: Use dimensional analysis to convert from grams to moles

$$0.035\text{ g KCl}\left(\frac{1\text{ mol KCl}}{74.55\text{ g KCl}}\right)$$

Step 3: Use dimensional analysis to continue to convert from moles to atoms

$$0.035\text{ g KCl}\left(\frac{1\text{ mol KCl}}{74.55\text{ g KCl}}\right)\left(\frac{6.02 \times 10^{23}\text{ atoms}}{1\text{ mol}}\right) = 2.8 \times 10^{20}\text{ atoms}$$

SUMMARY

The periodic table is made up of chemical elements which consist of protons, neutrons, and electrons. The three subatomic particles influence atomic number, atomic mass, and charge respectively. Don't forget that transition metals have more than one charge and be familiar with the ones that were pointed out in this chapter. This is particularly important in the chapter when we name molecules. Also be aware of the three trends of the periodic table and be prepared to answer multiple choice questions on them. Lastly, be comfortable calculating between moles and grams using molecular masses.

CHAPTER 14

PHASE DIAGRAMS AND CRYSTALLINE SOLIDS

OVERVIEW

For our intents and purposes, matter exists in three different phases: solids, liquids, and gases. In this chapter we will discuss two important phase diagrams. The first phase diagram examines the three phases in terms of pressure versus temperature. For this phase diagram we will discuss the important areas of the diagram that you will be responsible for. The second phase diagram that we will discuss is the heating curve. The important areas of this diagram will be pointed out, and an example question will be discussed regarding this diagram. After the phase diagrams we will examine the Clausius-Clapeyron equation which is an important equation that relates boiling temperatures and pressures. In the last part of this chapter we will talk about crystalline solids and their important aspects as well look at a as a sample question.

1. PHASE DIAGRAM FOR SOLIDS, LIQUIDS, GASES

The following diagram explains the three regions as the phase of matter that exists. It is important that you know which region belongs to which phase of matter. The regions will always be in order of solid, liquid, gas. Once you know which region corresponds to which phase of matter, the conversion from region to region should be obvious.

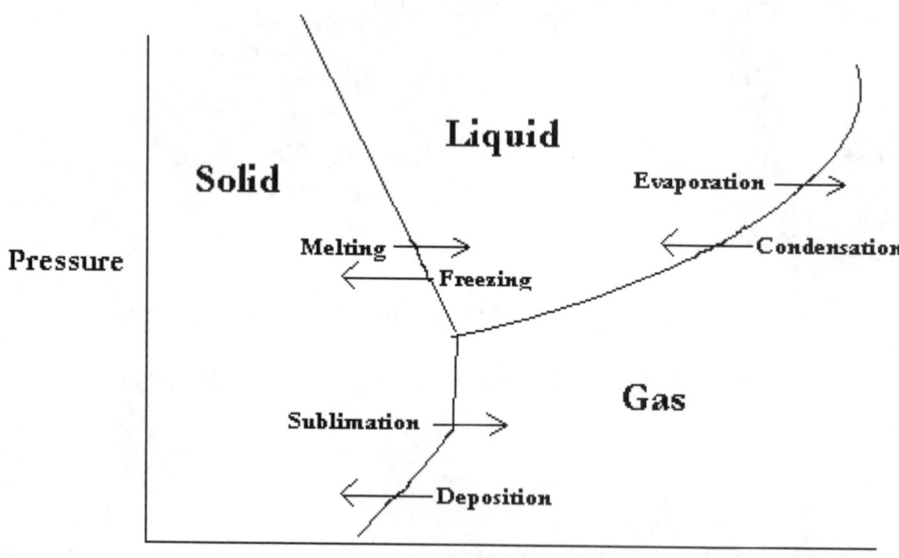

In the diagram the conversion from one phase of matter to another is summarized. You should be familiar with all of these phase changes because there is a good chance you will see them as multiple choice questions on an exam. As long as you are familiar with them, questions on changes between the phases of matter should be easy points.

Aside from knowing the phases of matter and the conversion between each phase, there are three very important points on the diagram that you will be expected to know. The triple point is the point in the middle of the diagram where all three boundaries meet. At this point solid, liquid, and gas all exist simultaneously. The next point that you will be responsible for is the normal boiling point. The normal boiling point occurs on the boundary between liquid and gas at 1 atmosphere pressure. The last important point is the critical point in which there is no boundary between liquid and gas. This occurs at the very top of the line between liquid and gas. All these are summarized below:

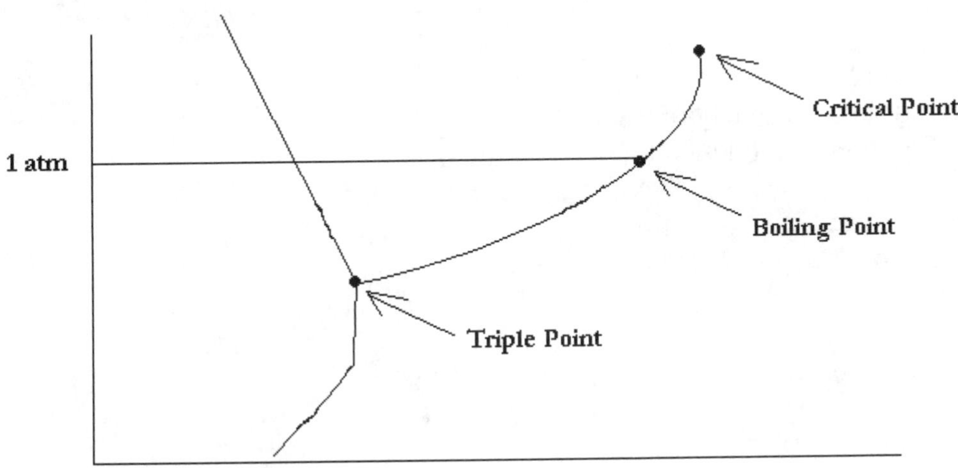

2. PHASE DIAGRAM OF WATER

As you heat H_2O from ice to steam, the diagram of temperature versus time is not a linear line. The reason for this is because as water changes from solid to gas, phase changes occur. As a phase change occurs, there is no change in temperature, so the line on the diagram stays flat. As a result, the phase diagram for water looks like a staircase. The following diagram is the phase diagram of water as it is heated over time:

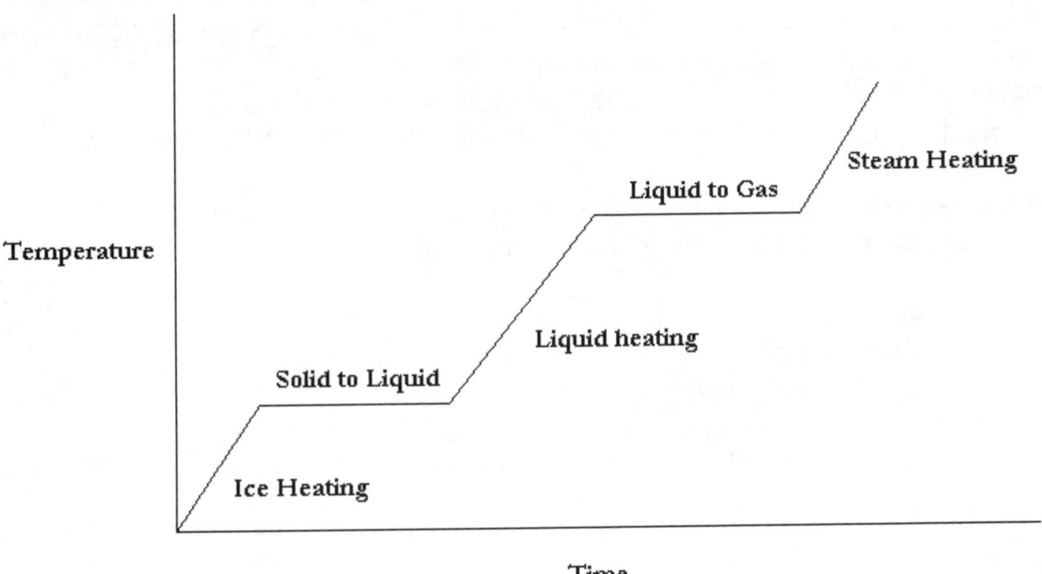

The typical question that you will see on an exam would ask you how much heat is necessary to bring water from one temperature to another temperature. The trick to these questions is that they will always involve at least one phase transition. When there is a phase transition involved, you cannot simply use one equation to solve the problem. Instead, you will have to combine two different equations.

Equation 1: Interval before or after phase transition
For an interval of temperature before a phase transition, use $q = mc\Delta T$.

where: q = heat required
m = amount of substance in moles
c = molar heat capacity of the sample
ΔT = Final temp – Initial temp

Equation 2: Phase transition
For the phase transition for solid to liquid, use $q = \Delta H_f \cdot n$
For the phase transition for liquid to gas, use $q = \Delta H_v \cdot n$

where: q = heat required
ΔH_f = enthalpy of fusion, or
ΔH_v = enthalpy of vaporization
n = amount of substance in moles

**Easy way of remembering this: The intervals before and after a phase transition are simply the regions on the diagram that are heating up. Therefore the only equation that involves a change in temperature is $q = mc\Delta T$. For the phase transitions, or the horizontal lines, there is no change in temperature and therefore you know you are not going to be using an equation that accounts for a change in temperature.

To solve these problems, I recommend first drawing a phase diagram of the substance. On the diagram I would mark the two temperatures that the question is asking about. Next, trace the line in the diagram between the two marks and ask yourself 'what kind of heating intervals am I dealing with'? Then use the appropriate equations for each interval on the diagram. Lastly add all the heats calculated and that will be the answer.

Example: Calculate the total heat that is required to raise 1mol of an unknown compound from 5° C to 100° C given the following information:

1. The compound melts at 5° C
2. The compound boils at 70° C
3. The enthalpy of vaporization is 40 kJ/mol
4. The molar heat capacity for the liquid phase is 60 J/K•mol
5. The molar heat capacity for the gas phase is 20 J/K•mol

Step 1: Draw heating curve for the unknown compound and mark the temperature interval that we are using

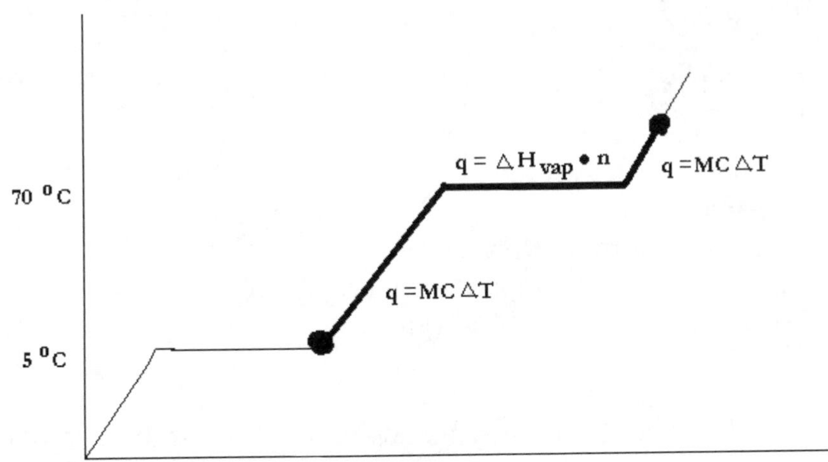

Step 2: Calculate heat for first interval (5 °C to 70 °C)

$q_1 = mc\Delta T$
$= (1 \text{ mol})(60 \text{ J/K}\cdot\text{mol})(70-5)$
$q_1 = 3900 \text{ J}$

Step 3: Calculate heat for the phase transition

$q_2 = \Delta H_v \cdot n$
$= (40 \text{ kJ/mol})(1 \text{ mol})$
$q_2 = 40 \text{ kJ} = 40{,}000 \text{ J}$

Step 4: Calculate heat for the last interval (70°C to 100°C)

$Q_3 = mc\Delta T$
$= (1 \text{ mol})(20 \text{ J/K}\cdot\text{mol})(100-70)$
$q_1 = 600 \text{ J}$

Step 5: Add all of the heats together to determine the total heat required. Be sure to use units of J for the second heat:

$q_T = q_1 + q_2 + q_3$
$= 3900 \text{ J} + 40{,}000 \text{ J} + 600 \text{ J}$
$q_T = 44{,}500 \text{ J}$

**Note that almost everyone gets points taken off for this problem because they forget to convert the second heat to J. We do this because of the units in the given information in the problem are in J except for the enthalpy of vaporization. You cannot add two values in joules to one value in kilojoules, so be careful to convert first!

3. CLAUSIUS-CLAPEYRON EQUATION

The Clausius-Clapeyron equation is an important means of relating boiling temperatures and pressures.

Definition: $\ln\frac{P_2}{P_1} = \frac{\Delta H_{vap}}{R}(\frac{1}{T_1} - \frac{1}{T_2})$

Components:
P = Pressure
T = Temperature (in Kelvin)
ΔH_{vap} = Enthalpy of vaporization (in kJ/mol)
R = Gas constant (8.314 J/K•mol)

There are three important things that you must be aware of. The first thing to notice is that the value gas constant is NOT the one we are used to using. Instead of using 0.0821 L•atm/mol•K, we must use 8.314 J/K•mol. This is an important value and you must memorize this value and the units and not confuse the two.

Another important thing that you must be cautious about is ΔH_{vap}. This value is usually given in kJ/mol. However, the gas constant uses Joules. In order for the units to cancel properly, you MUST first convert ΔH_{vap} into Joules. Failure to do so will lead you to an incorrect answer.

The last important thing that you must be aware of is keeping track of the variables. You must be consistent when assigning variables to the subscripts 1 and 2. My best advice is to always assign the subscript 2 variables as the boiling conditions and the subscript 1 variables as the current conditions. This way you will keep everything consistent and eliminate the possibility of getting these questions wrong.

Example: Calculate the boiling point of propanol if the vapor pressure at 60° C is 300 torr and the enthalpy of vaporization is 40.5 kJ/mol.

Step 1: Recognize that this problem calls for the Clausius-Clapeyron equation because you are asked for a boiling point given a temperature and pressure along with an enthalpy of vaporization.

Step 2: Write the Clausius-Claperyron equation

$$\ln\frac{P_2}{P_1} = \frac{\Delta H_{vap}}{R}(\frac{1}{T_1} - \frac{1}{T_2})$$

Step 3: Determine each variable by assigning all of the subscripts of 1 as the current condition and the subscripts of 2 as the boiling point. Remember that boiling occurs at 760 mm torr, or 1 atm. Also, don't forget to convert °C to K by adding 273:

P_1 = 300 torr
P_2 = 760 torr (which is 1 atm)
T_1 = 60 + 273 = 333
T_2 = ?

Step 4: Plug variables into equation and use algebra to solve for T_2

$$\ln\frac{760}{300}=\frac{40500}{8.314}\left(\frac{1}{333}-\frac{1}{T_2}\right)$$

$$0.930=4871\left(0.003-\frac{1}{T_2}\right)$$

$$0.930=14.61-4871\left(\frac{1}{T_2}\right)$$

$$-13.68=-4871\left(\frac{1}{T_2}\right)$$

$$0.0028=\frac{1}{T_2} \text{ so } T_2=357\text{ K}$$

4. CRYSTALLINE SOLIDS

There are three cubic structures of crystalline solids that you will have to differentiate. The key part of crystalline solids is to first recognize which type of cubic structure the question is dealing with, and then you will know which equation to use. Each type of cubic structure has its own formula for radius and volume and each is illustrated below.

4a. Primitive Cubic (or Simple Cubic)

Image: 

Number of atoms = 1
Edge = 2r
Volume = $(2r)^3$

4b. Face Centered Cubic

Image: 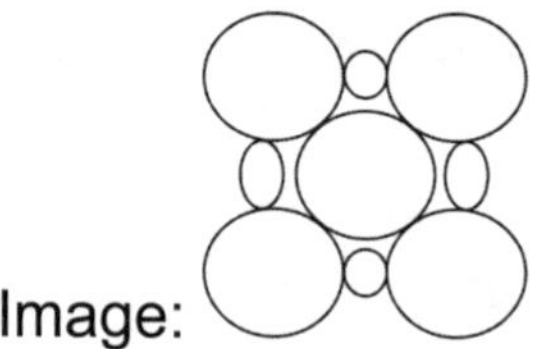

Number of atoms = 4

Edge = $\dfrac{4r}{\sqrt{2}}$

Volume = $\left(\dfrac{4r}{\sqrt{2}}\right)^3$

4c. Body Centered Cubic

Image: 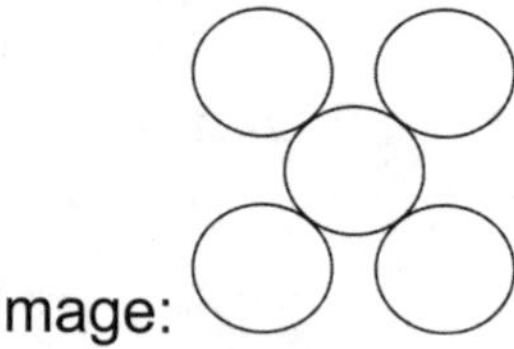

Number of atoms = 2

Edge = $\dfrac{4r}{\sqrt{3}}$

Volume = $\left(\dfrac{4r}{\sqrt{3}}\right)^3$

The formulas for crystalline structures seem very complicated, however, when you think about it they are not that terrible. Face-centered and body-centered have close to identical formulas. They only differ by the denominator.

**Easy way of remembering the formulas: In order to remember that face centered has the denominator of 2 and not 3, think of a human face. A human face has 2 eyes, 2 ears, 2 nostrils. Therefore, think of the number 2 as face centered cubic and you will never confuse face centered and body centered formulas.

There are two units that are often used for questions involving crystalline structures that you have to be familiar with. A pico-meter is 10^{-12} meter and is often the unit used to express an atomic radius. Make sure you know this conversion factor between picometers and meters. Another important unit that you will need to know is Angstroms. An Angstrom is 1.0×10^{-10} meter. Angstroms are denoted by the symbol Å: an A with a circle above it. Be sure to know this conversion because you will likely not be given the conversion on an exam.

The typical question that you will be asked for crystalline structures is to calculate the density of an atom. The way that you should handle a question like this would be to first calculate the volume of the atom using the appropriate formula based on the edge or radius that is given in the problem. Next you would want to calculate the weight of the atom using the molecular mass and the number of atoms in that particular cubic structure. Lastly, you would need to divide the weight by the volume and you will be left with density.

Example: Calculate the density of copper using a Face Centered Cubic structure. The radius of a copper atom is 130 pm and the molar mass is 63.5 g/mol.

Step 1: Write the formula for the volume of a face centered cubic structure:

$$\text{Volume} = \left(\frac{4r}{\sqrt{2}}\right)^3$$

Step 2: Calculate volume by plugging the radius into equation. Be sure to convert pm to cubic centimeters with dimensional analysis:

$$V = \frac{4}{\sqrt{2}}\left[(130\,\text{pm})^3\left(\frac{1\,\text{m}}{10^{12}\,\text{pm}}\right)^3\left(\frac{100\,\text{cm}}{1\,\text{m}}\right)^3\right] = 6.21\times10^{-24}\,\text{cm}^3$$

Step 3: Calculate the weight using the fact that face centered cubic structures have 4 atoms:

$$4\,\text{atoms}\left(\frac{1\,\text{mol}}{6.02\times10^{23}\,\text{atoms}}\right)\left(\frac{63.5\,\text{g}}{1\,\text{mol}}\right) = 4.22\times10^{-22}\,\text{g}$$

Step 4: Calculate density using volume and mass calculated above:

$$D = \frac{m}{v}$$

$$= \frac{4.22\times10^{-22}\,\text{g}}{6.21\times10^{-24}\,\text{cm}^3}$$

$$D = 68\,\tfrac{\text{g}}{\text{mL}}$$

SUMMARY

In this chapter we discussed two important phase change diagrams; talked about the Clausius-Clapeyron equation and its applications, and looked at crystalline structures. For the two phase diagrams make sure you understand the diagrams, and know how to interpret information from them. For the Clausius-Clapeyron equation, be sure to be consistent with the variables and assign the same subscripts the same conditions. For crystalline structures, make sure you identify which cubic structure the question is asking for and use the proper equation. For each cubic structure it is important that you know the number of atoms and the formula for volume.

CHAPTER 15

QUANTITATIVE CALCULATIONS

OVERVIEW

In this chapter we will discuss percent composition, empirical and molecular formulas, balancing equations, stoichiometric ratios, limiting reagents, theoretical yield and hydrates. This chapter will guide you through how to recognize the particular type of question when it appears on an exam, how to go about solving them, and how to use judgment on whether or not the answer makes sense.

1. PERCENT COMPOSITION

Percent composition questions are easy to recognize because they always start with the same phrase, "Find the percent of x, y, and z in the following compound…" The way that you should handle these types of questions is by viewing every compound as a group of individual elements. Each element contributes a certain amount to the overall compound. Think of it in terms of parts over total. Ask yourself how much does this particular element's mass account for the overall mass of the compound? The way that you should do this is by first finding the overall mass of the compound. Then divide the mass of each element by the mass of the overall compound and multiply this by 100.

Example: Find the percent of C, H, and O in $C_6H_{12}O_6$

Step 1: Find the total mass of the compound, using the atomic masses of each element

6 × C: 12.01 × 6 = 72.06
12 × H: 1.008 × 12 = 12.096
6 × O: 16 × 6 = 96
Total = 72.06 + 12.096 + 96 = 180.156 g

Step 2: Find the percent of each element by dividing the mass of the element by the total mass of the compound

$$\%C = \frac{72.06}{180.156} \times 100 = 40\,\%$$

$$\%\,H = \frac{12.096}{180.156} \times 100 = 6.7\,\%$$

$$\%O = \frac{96}{180.156} \times 100 = 53.3\,\%$$

Step 3: Check to see if your percents make sense

40 + 6.7 + 53.3 = 100 %

Keep in mind that the percents of each element should sum to 100 because they all combine to form one compound. Also bigger elements, and the elements that appear the most, are going to be the elements that should have the highest percent mass. If you have small elements with high percents, or percents that add to way under or over 100, make sure you check your math.

2. EMPIRICAL AND MOLECULAR FORMULAS

A chemical formula shows the specific ratio of elements that combine to form a compound. If the formula shows the smallest whole number ratios that cannot be reduced it is known as an empirical formula. An example of an empirical formula would be CH_4. A molecular formula is a chemical formula that can be reduced further and is a multiple of an empirical formula. An example of a molecular formula would be C_2H_8. C_2H_8 is a molecular formula since both of the numbers are divisible by 2 and can be reduced. C_2H_8 is two times the empirical formula CH_4.

The typical questions that are asked about empirical and molecular formulas start with a given percent of each element in a compound and its molecular mass; and you then have to find the empirical formula of the compound. Do not be fooled if the question just flat out asks for a chemical formula. Regardless of how they word the question, you will always solve it the same way by first finding the empirical formula, as explained below.

The most important thing to remember to do is to assume that we have 100 g of compound; and then we can convert the percents directly grams. Next we convert all of these grams to moles so that we can determine the ratio between all of the elements in the empirical formula. To determine the mole ratios, divide each value of moles by the smallest number of moles in the list. We might need to round up a little to the nearest integer. At this point, we can write the empirical formula directly using these mole ratios. However, since we want the molecular formula the final step is to calculate the mass of the empirical formula and divide the molecular mass by this value. We then multiply every subscript in the empirical formula by this value.

Example: A compound with a mass of 120 g is 40 % Carbon, 6.7 % Hydrogen and 53.3 % Oxygen. What is the chemical formula of the compound?

Step 1: Calculate the moles of each element by taking the percentages as grams:

$$40\ \text{g C}\left(\frac{1\ \text{mol C}}{12.01\ \text{g C}}\right) = 3.33\ \text{mol C}$$

$$6.7\ \text{g H}\left(\frac{1\ \text{mol H}}{1.008\ \text{g H}}\right) = 6.65\ \text{mol H}$$

$$53.3\ \text{g O}\left(\frac{1\ \text{mol O}}{16\ \text{g O}}\right) = 3.33\ \text{mol O}$$

Step 2: Determine the ratio of each element by dividing each number of moles by the smallest number (which in this case is 3.33). Round each to the nearest integer:

C	H	O
$\frac{3.33\ \text{mol}}{3.33\ \text{mol}} = 1\ \text{mol C}$	$\frac{6.35\ \text{mol}}{3.33\ \text{mol}} = 1.9\ \text{mol H} \approx 2\ \text{mol H}$	$\frac{3.33\ \text{mol}}{3.33\ \text{mol}} = 1\ \text{mol}$
1	2	1

Step 3: Write the empirical formula, using these ratios as subscripts:

$C_1H_2O_1 = CH_2O$

Step 4: Determine the mass of the empirical formula:

1 × C: 12.01 × 1 = 12.01
2 × H: 1.008 × 2 = 2.016
1 × O: 16 × 1 = 16
Total = 12.01+ 2.016 + 16 = 30.036 g ≈ 30 g

Step 5: Determine the multiple of the empirical formula by dividing the mass of the molecular formula by the mass of the empirical formula.

$$\frac{120\ \text{g}}{30\ \text{g}} = 4$$

Step 6: Multiply each element in the empirical formula by 4 to get the molecular formula

$C_4H_8O_4$

Step 7: Check your work by making sure the mass of this compound adds to ≈ 120 g:

4 × C: 12.01 × 4 = 48.04
8 × H: 1.008 × 8 = 8.064
4 × O: 16 × 4 = 64
Total = 48.04 + 8.064 + 64 ≈ 120 g

3. BALANCING EQUATIONS

Being able to balance chemical equations is a vital skill that will be necessary for the successful completion of a lot of questions in chemistry. It is very important that you balance the equations correctly, because your answer will directly depend on a balanced equation.

My best advice for balancing equations is to first look at both sides of the equation and keep in mind what elements are on each side. The elements that are present the least amount of times on both sides are the elements that you want to start with. If you think about it, you do not want to start with elements that appear more than once on either side of the equation because the balancing will constantly be offset once you try to start balancing other elements. The way that I look at it is that you want to start with the elements that will be least affected by balancing the rest of the equation and then start to focus on the more troublesome elements. From my experience I have noticed that Oxygen tends to be the most troublesome element to balance and in most cases you are going to want to leave it until the end. Another word of advice is that you are going to want to leave elements that are by themselves until the very end.

Example: Balance _____C_3H_8 + _____O_2 → _____CO_2 + _____H_2O

The first element that we should want to balance is carbon, since carbon is present only once on either side of the equation. Then, as there are 3 carbons on the left side of the equation, we can put a coefficient of 3 in front of the carbon containing compound on the right.

Next, Hydrogen is also an element that exists only once on either side of the equation. Hydrogen appears 8 times on the left side but only 2 times on the right. Therefore, try a coefficient of 4 on the H_2O in order to make it have 8 hydrogens.

Lastly, we are left with oxygen because it is an element that is by itself and so we should leave it for last. If you add all of the oxygen elements on the right you will notice that there are a total of 10. In order to get 10 oxygen elements on the left side, you will have to but a coefficient of 5 in front of O_2 on the left:

Answer: _____C_3H_8 + __5__ O_2 → __3__ CO_2 + __4__ H_2O

As you can see, if you didn't leave O_2 for last, by balancing the other elements would have disrupted the balancing for oxygen and the entire equation would be incorrectly balanced.

4. STOICHIOMETRIC RATIOS

A stoichiometric ratio is very simply the ratio of coefficients of any two compounds in a chemical equation. These ratios are important because they are used in dimensional analysis problems involving chemical reactions in order to switch from a reactant to a product. As you will see in the next section, this is particularly important when doing limiting reagent questions. Before we get to limiting reagents, let's first examine stoichiometric ratios and how they are used.

Example: How many grams of CO_2 can be produced when 7.0 g of O_2 react with excess C_3H_8 in the following reaction?

_____C_3H_8 + __5__ O_2 → __3__ CO_2 + __4__ H_2O

Step 1: Since we will use the sotichiometric ratios to get this answer, we first need to convert 7.0 g O_2 to mol O_2:

$$7.0\,\mathrm{g\,O_2}\left(\frac{1\,\mathrm{mol\,O_2}}{32\,\mathrm{g\,O_2}}\right)$$

Step 2: Use stoichiometric ratios to convert from O_2 (reactant) to CO_2 (product)

$$7.0\,\mathrm{g\,O_2}\left(\frac{1\,\mathrm{mol\,O_2}}{32\,\mathrm{g\,O_2}}\right)\left(\frac{3\,\mathrm{mol\,CO_2}}{5\,\mathrm{mol\,O_2}}\right)$$

Step 3: Convert from moles to grams of CO_2 since that is what the question is asking for

$$7.0\,\mathrm{g\,O_2}\left(\frac{1\,\mathrm{mol\,O_2}}{32\,\mathrm{g\,O_2}}\right)\left(\frac{3\,\mathrm{mol\,CO_2}}{5\,\mathrm{mol\,O_2}}\right)\left(\frac{44\,\mathrm{g\,CO_2}}{1\,\mathrm{mol\,CO_2}}\right)=5.8\,\mathrm{g\,CO_2}$$

5. LIMITING REAGENT

As you saw in the previous example, there was only one value given for one of the reactants while the other reactant was said to be in excess. In Limiting Reagent questions, also known as limiting reactant questions, there will be a starting amount given for each reactant, and we then have to determine which one is limiting. The general rule of thumb is that you can only make as much product as the least amount of reactant that you start with in a chemical reaction. So in limiting reactant questions you need to calculate the amount of product twice, starting each time from the two different amounts of reactant given. Then whichever reactant gives the least amount of product will be the limiting reactant.

Before we get into how to figure out which one is the limiting reactant, first let's examine what is meant by the phrase "you can only make as much product as the least amount of reactant." Let's say you are baking a cake. The recipe of each cake calls for 1 box of cake mix and 1 cup of milk. If you had 20 cake mixes, but only 2 cups of milk, you can still only make 2 cakes. In this particular example, the milk is the limiting reactant. Once we use the 2 cups of milk the milk gets entirely used up and we can no longer make any cake. Even though we still have an excess of 18 cake mixes, there is nothing we can do with them and they are left in excess. This relates back to chemistry because there will always be one reactant that gets used up faster than the other reactant. When this happens, the reaction will stop and no more products can be produced.

Questions on limiting reactants will either ask you to simply identify which reactant is limiting, or may also ask you to calculate how much product is formed in a limiting reactant situation. This second option is actually very easy. Since you will have to do this calculation in any case to figure out the limiting reactant, you just take the amount of product that the limiting reactant formed as the answer to how much product is formed.

In either case, when there is a starting amount given for each reactant you must automatically recognize the question as a limiting reactant question. Once you have done this, calculate the amount of product formed from each reactant using the steps outlined in section 4.

Example: How many grams of CO_2 can be produced when 5.0 g of O_2 react with 10.0 g C_3H_8 in the following reaction?

_____ C_3H_8 + ___5___ O_2 → ___3___ CO_2 + ___4___ H_2O

Step 1: Recognize this is a limiting reactant question, since it gives amounts of two reactants. Start by calculating how many g CO_2 are formed from each reactant separately in turn, using the steps for stoichiometric ratio calculations:

$$5.0\,g\,O_2\left(\frac{1\,mol\,O_2}{32\,g\,O_2}\right)\left(\frac{3\,mol\,CO_2}{5\,mol\,O_2}\right)\left(\frac{44\,g\,CO_2}{1\,mol\,CO_2}\right)=4.1\,g\,CO_2$$

$$10.0\,g\,C_3H_8\left(\frac{1\,mol\,C_3H_8}{44.11\,g\,C_3H_8}\right)\left(\frac{3\,mol\,CO_2}{1\,mol\,C_3H_8}\right)\left(\frac{44\,g\,CO_2}{1\,mol\,CO_2}\right)=29.9\,g\,CO_2$$

Since O_2 gives the least amount of CO_2, it is the limiting reactant.

Step 2: If we know that O2 is limiting, we already know that 4.1 g will be produced.

Answer: 4.1 g CO_2

6. THEORETICAL YIELD

Theoretical yield is the answer that you obtain after doing a dimensional analysis calculation to figure out how much product is formed. In other words, it is theoretically the maximum amount of product you could get from the reaction in the ideal world. However, as we know, this is not an ideal world. The amount of product that was calculated as your theoretical yield is typically not the actual amount of product that you are left with at the end of the experiment, since in practice you would always get less. The amount that you are actually left with is known as your actual yield.

In the laboratory, it is very important for scientists to know how close their actual yield was to their theoretical maximum yield that they should have obtained. In order to figure this out, we use a formula known as percent yield. The equation for % yield is:

$$\%\ \text{yield} = \frac{\text{Actual yield}}{\text{Theoretical yield}} \times 100$$

A lot of people confuse the terms actual and theoretical yield. Think about it rationally. The actual yield is the amount that was ACTUALLY obtained. This value is usually given in the question and you are often asked for the percent yield by using that value along with the value that you just calculated in the question. The theoretical yield, the answer to your calculation, is the amount that THEORETICALLY should be obtained in an ideal perfect world.

7. HYDRATES

A hydrate is a salt compound that has water molecules attached to the structure. The typical question about hydrates asks you to find the coefficient in front of the water molecule that is attached to the salt. In order to find this coefficient, you will need to find the ratio between the moles of salt and the moles of water. The way that this can be done is by being given the mass of the hydrate before and after being heated, which will drive off the water. The difference between the two masses is equal to the mass of the water since that's how much mass was lost after the water evaporated. From here you can very simply calculate the moles of salt and moles water and determine the ratio.

Example: When 35.2 g of the hydrate CoCl2•xH_2O was heated to remove all the water, the remaining salt weighed 20.8 g. What is the value of the coefficient x?

Step 1: Write the chemical equation

$CoCl_2 \cdot xH_2O \rightarrow CoCl_2 + xH_2O$

Step 2: Plug in the information given

$CoCl_2 \cdot xH_2O \rightarrow CoCl_2 + xH_2O$
35.2g 20.8g 14.4g
(this value is obtained by subtracting 35.2 – 20.8)

Step 3: Calculate the moles of the salt and the water

$$20.8\,\text{g CoCl}_2\left(\frac{1\,\text{mol CoCl}_2}{129\,\text{g CoCl}_2}\right) = 0.161\,\text{mol CoCl}_2$$

$$14.4\,\text{g H}_2\text{O}\left(\frac{1\,\text{mol H}_2\text{O}}{18\,\text{g H}_2\text{O}}\right) = 0.8\,\text{mol H}_2\text{O}$$

Step 4: Determine the ratio between the salt and the water

$$0\frac{0.8}{.161} \approx 5$$

Answer: This ratio tells us there must be 5 moles of water in the hydrate:

Chemical formula: $CoCl_2 \cdot 5H_2O$

SUMMARY

In this chapter we covered percent composition, empirical and molecular formulas, balancing equations, stoichiometric ratios, limiting reagents, theoretical yield and hydrates. It seems like a lot of material, but rest assured most of the math that you will be doing in introductory chemistry resides in this chapter and stays in this chapter. This chapter illustrated the different quantitative calculations as well as told you the exact ways that these questions are usually asked. You should notice that all of these questions can be handled in a very systematic way, going step by step and arriving at an answer. Do not be scared by the wording of a question and as this chapter should have demonstrated it is not nearly as hard as it looks.

My best advice would be to first recognize the type of question that is being asked. Ask yourself things such as “Are they giving me a starting mass for both reactants?” (so it is a limiting reagent) or questions like “Does this question deal with a salt and water complex?” (hydrate problem). Once you recognize the question, go about the systematic approach as I pointed out in this chapter. As long as you are careful to not make any mathematical mistakes or calculator errors, there should be no reason why you can’t get these types of questions fully correct. Remember also to use dimensional analysis where ever possible.

CHAPTER 16

SOLUBILITY AND NET IONIC EQUATIONS

OVERVIEW

Solubility is a subject that has to do with what can and cannot dissociate or break apart in water. One main purpose of studying solubility is to be able to determine the chemical makeup of a precipitate at the bottom of a test tube after mixing two solutions together. A precipitate is the insoluble solid that forms as a result of mixing two compounds. In this chapter we will first discuss the important solubility rules. We will then use those rules to examine the important concepts that require the use of solubility rules. At the same time we will examine the types of questions that are usually asked on tests and quizzes relating to this topic.

1. SOLUBILITY RULES

1. Any NO_3^-, $C_2H_3O_2^-$, HCO_3^-, ClO_3^-, ClO_4^- salts are always soluble
2. NH_4^+ salts are always soluble
3. Salts of Group 1 metals (Li, Na, K etc) are always soluble
4. Cl^-, Br^-, I^- salts are soluble, except if they have Ag^+, Hg_2^{+2}, and Pb^{+2}
5. $SO4^{-2}$ salts are soluble except if they have Ag^+, Hg_2^{+2}, Pb^{+2}, Ca^{+2}, Sr^{+2}, Ba^{+2}
6. CO_3^{-2}, CrO_4^{-2}, $C_2O_4^{-2}$, PO_4^{-3} salts are are insoluble except when bonded to NH_4^+ or any group 1 metal
7. S^{-2}, OH^- salts are insoluble except when bonded to NH_4^+, any group 1 metal, Ca^{+2}, Sr^{+2}, Ba^{+2}

My advice for memorizing the solubility rules is to pick up on generalizations. As you can see, there seems to be a lot of rules that you need to memorize as well as specific exceptions. What I recommend is to learn these solubility rules by making generalizations such as, "whenever I see a group 1 metal I know the compound is soluble no matter what". Or "If ever I see an NH_4^+ I know the compound is going to be soluble". Also things such as "Whenever I see cations like Ca^{+2}, Sr^{+2}, and Ba^{+2} I should be suspicious because these cations are often seen as exceptions"

As you will see when you continue reading this chapter, it is important to recognize the solubility rules more so then strictly memorizing them. As long as you are comfortable enough to recognize them, you should be in pretty good shape. For the most part, every compound is going to have two rules associated with it, one for the cation and one for the anion. If you can at least remember the rule for one of the two you will be able to identify whether or not the overall compound is soluble or insoluble.

For example:

$NaNO_3$ = Soluble

You can recognize this by either remembering that all group 1 metals are soluble, so this compound has to be soluble. But if you didn't remember that particular rule you might remember that NO_3 is always soluble; therefore this compound has to be soluble since it has NO_3. As long as you can recognize at least one of the rules associated with the salt then you should be able to determine whether or not it is soluble.

2. STRONG VS. WEAK ELECTROLYTES

Typically, instructors don't usually flat out ask a question such as "Is the following molecule soluble or insoluble?" Instead they will ask a question about solubility in disguise. So they might set a multiple choice question that asks "Which of the following is a strong electrolyte?"

What is meant by the term strong electrolyte is a substance that has the ability to break apart into its respective ions completely. In other words a strong electrolyte is a compound that is soluble. Strong acids and strong bases are also strong electrolytes because they completely dissociate and break apart. Therefore, whenever you are asked a question about a strong electrolyte you are looking for an answer that is soluble, strong acid, or strong base.

A weak electrolyte is a substance that is soluble but that does not have the ability to break apart into ions easily. Often weak acids and weak bases are also weak electrolytes because they do not completely dissociate in water. Therefore, whenever you are asked a question about a weak electrolyte you are looking for an answer that is a weak acid, or weak base.

Example: Which of the following compounds is NOT a weak electrolyte?
A) AgCl   B) $BaSO_4$ C) NaCl   D) Hg_2SO_4

Answer: choice C.

Explanation: Remember, we are asked which is NOT a weak electrolyte; meaning which one of the choices is a strong electrolyte. Do not get tricked by the wording and be sure to read the question carefully. NaCl is a strong electrolyte because it is soluble. You can recognize that it is soluble two different ways. The first being that it contains a group 1 metal, the second being that it contains a halogen which is soluble and Na is not one of the exceptions.

3. DOUBLE DISPLACEMENT REACTIONS (METATHESIS REACTIONS)

The most common types of questions about solubility often focus on a double displacement reaction which is also known as a metathesis reaction. Again, this is a very scary sounding name that is not that hard. A double displacement reaction occurs with the mixing of two salt solutions which leads to two different products. When two salts are mixed together, the positive ion of the first salt bonds with the negative ion of the second salt to make one product. Also, the negative ion of the first salt then bonds with the positive ion of the second salt to make a second product. Keep in mind when writing the ionic compounds that the positive ion comes first in th chemical formula

An example of a double displacement reaction:

$$NaCl + AgNO_3 \longrightarrow NaNO_3 + AgCl$$

Once you have performed a double displacement reaction, you will be left with two new products. One of the products will almost always be soluble and one will almost always be insoluble. You must be able to determine which one is soluble and which one is insoluble based on the solubility rules. If you remember from our previous example, $NaNO_3$ is soluble. AgCl is insoluble because Cl is soluble but one of the exceptions is Ag therefore making this compound insoluble.

Now that we have determined which compound is soluble and which one is not, we can answer any other type of question that the instructors might ask. Examples of these other types of questions that you can be asked are explained in the next two sections.

3a. Net ionic equation

The net ionic equation is the essence of a double displacement reaction that forms a precipitate. The only question we really care about is what is the chemical make-up of that precipitate at the bottom of the test tube and why did it form. In order to answer this question, we write what is called a net ionic equation. In the net ionic equation we focus only on the two ions that formed the insoluble compound and ignore everything else.

The net ionic equation of the above example is:

$$Ag^+ + Cl^- \longrightarrow AgCl\ (s)$$

3b. Spectator ions

The ions that are ignored in the net ionic equation are called spectator ions. The way that I think of spectator ions is that they are just there to watch – just as if you go to a baseball game and sit in the stands you are considered a spectator. You do not participate in the baseball game; you simply are there to watch the show. This is exactly what happens with spectator ions. They are the ions that form the soluble compound and have nothing to do with the insoluble compound. Therefore, these ions are just along for the ride.

The spectator ions in the original example would be:

Na^+ and NO_3^-

SUMMARY

Solubility simply refers to a compound's ability to break apart. In this chapter we discussed the rules for determining solubility as well as looked at the ways in which solubility questions will be asked on a test or quiz. Make sure you are comfortable with the terms strong and weak electrolytes. Also be comfortable with a double displacement reaction and be prepared to answer the questions that can be asked about them. Remember the questions that will follow a double displacement reaction will be questions such as, "Which of the following products is soluble/insoluble?", "Which of the following ions is a spectator ion?", or "Write the net ionic equation".

CHAPTER 17

STATISTICS

OVERVIEW

Statistics has become increasingly popular in general chemistry and is an important tool for data analysis. Usually you will not be doing any complicated statistics, but you will be responsible for a few important statistical equations. Statistics also becomes important in chemistry for analyzing results obtained in the lab. The way that the results are analyzed is by comparing them to one another to determine their precision; to see if there is a data point that can be statistically discarded, or to determine how confident you can be with data obtained from an experiment. All of these can be determined by the statistics of standard deviation, the Q test, and confidence intervals.

1. STANDARD DEVIATION

Equation: $SD = \sqrt{\frac{\sum (X-\overline{X})^2}{n-1}}$

Components: X = Value
$\overline{X}$ = Mean
n = number of values

You should think of standard deviation as a measure of dispersion around the mean. Standard deviation is basically a value that tells us how much each of a set of readings stray from their mean value. The equation for standard deviation may look extremely complex and difficult: however, it is very easy. The symbol '∑' is a Greek letter that means "the sum of". In this equation the ∑ symbol is telling us to subtract the mean from each number in the list, square each result, then add all of those values together.

In other words, we need to evaluate $(X_1 - \overline{X})^2 + (X_2 - \overline{X})^2 + (X_3 - \overline{X})^2$...for as many numbers are in the list

In order to calculate the standard deviation of a set of numbers:

1. First calculate the mean. To do this, simply add all of the numbers together and divide that total by the amount of numbers in the list
2. Take the first number in the list and subtract the mean from it
3. Square the quantity from step 2
4. Repeat steps 2 and 3 for each remaining number in the list
5. Add all of the values together that were obtained in step 3 for all the numbers
6. Divide the quantity obtained in step 5 by one less than the amount of numbers
7. Take quare root the value obtained in step 6

Example: Calculate the standard deviation for the following list of five numbers: 2.01, 2.93, 1.99, 3.04, 2.86

Step 1: Calculate the mean:

2.01 + 2.93 + 1.99 + 3.04 + 2.86 = 12.83;

$$\frac{12.83}{5} = 2.57$$

Step 2: Take each number from the list, subtract the mean from it, square the value, and then add them all together:

$$(2.01-2.57)^2 + (2.93-2.57)^2 + (1.99-2.57)^2 + (3.04-2.57)^2 + (2.86-2.57)^2 = 1.08$$

Step 3: Divide this quantity by the amount of numbers (5) minus 1 (in other words 4):

$$\frac{1.08}{4} = 0.27$$

Step 4: Take the square root:

$$\sqrt{0.27} = 0.52$$

So the standard deviation is 0.52, and we express how the numbers stray from the mean as 2.57 ± 0.52

2. Q TEST

Equation: $Q = \dfrac{|SuspiciousValue - ClosestValue|}{HighestValue - LowestValue}$

The idea of performing a Q test is to determine if we can discard a data point that was collected. If there is a data point that is an extreme outlier compared to the rest, this point will cause us to have an extremely bad standard deviation. Instead, we perform a Q test to determine if this is statistically significant or if we can throw the point out.

The way that we perform a Q test is by using the equation above. We take the outlier and subtract it by the next closest data point in the list. Then we divide that quantity by the range of data points. The range is simply the largest value minus the smallest value in the list of numbers. This value that you obtain from the equation must then be compared to a table of values. The table of values is a list of critical Q values for varying amounts of numbers in the list. In other words the amount of numbers in our list will determine which column we should look to in the table to compare our calculated Q to the critical Q.

If the measured Q is greater than the critical Q, then the data point can be excluded.

If the measured Q is less than the critical Q, then the data point can NOT be excluded.

Keep in mind that a question on a test might not ask you to perform a Q test. Instead, it might be disguised as a standard deviation question. So the question may ask you to calculate the standard deviation of a list of numbers. However, you will notice that there is one data point that is an extreme outlier in the list. If an outlier exists, then you should first perform a Q test to see if you can neglect that value or not. Then, once you have performed the Q test, you should then go ahead and calculate your standard deviation.

Example: Can any of the data be statistically rejected from the list below?

5.23, 5.44, 5.38, 5.69, 10.88, 5.21

Step 1: Recognize that 10.88 is an apparent outlier in this list of numbers

Step 2: Plug in the values into the equation:

$$Q = \frac{|10.88 - 5.69|}{10.88 - 5.21}$$

Q= 0.915

Step 3: Compare calculated Q to the critical Q in table (for 6 values):

N	3	4	5	6	7	8
Q_c	0.94	0.76	0.64	**0.56**	0.51	0.47

Calculated Q = 0.915
Critical Q = 0.56
Or Calculated Q > Critical Q

Step 4: Make the determination:

Since the calculated Q is greater than the critical Q, the data point CAN be excluded

3. CONFIDENCE INTERVAL

Equation: $C = \overline{X} \pm \frac{TS}{\sqrt{n}}$

Components: $\overline{X}$ = mean
T = tabulated value
S = standard deviation
n = amount of numbers

A confidence interval is exactly what the name implies. It is a range of numbers for which we can have a certain amount of confidence in our number if it falls within this range. Confidence intervals are usually calculated for a 95 % confidence level. So if a number falls within the calculated range, we can be 95 % confident in the number obtained. An alternative way to think of this is that if we performed our experiment many times, then this value would be reliable and accurate 95 % of the time. To calculate a confidence interval you will need to have first calculated the standard deviation. With the standard deviation and the tabulated value, you can determine the confidence interval.

Example: Determine the 95 % confidence interval for the data obtained below:

2.01, 2.93, 1.99, 3.04, 2.86

Step 1: Calculate the standard deviation

0.52
(This was the standard deviation example in the previous section!)

Step 2: Determine the tabulated value (t)

	80%	90%	95%	99%
1				
2				
3				
4			**2.78**	
5				

Step 3: Plug values into equation

$$C = 2.57 \pm \frac{(2.78)(.52)}{\sqrt{5}}$$

$$2.57 \pm .647$$

Step 4: Determine confidence interval. Subtract 0.647 from the mean to determine the beginning of the interval and add 0.647 to the mean to determine the end of the interval:

95 % confidence interval = 1.923 - 3.217

SUMMARY

Statistics is an important means of analyzing data that we obtain in chemistry. The three major areas of statistics that you will be tested on in introductory chemistry are standard deviation, Q tests, and confidence intervals. When we obtain data, we must be sure that we know what the data means and how strongly we can have confidence in our data. In a situation where a data point is an outlier, we cannot simply get rid of the number on the basis of our own discretion. Therefore, there must be a systematic way of handling outliers and see whether they are statistically significant or if they can be rejected.

CHAPTER 18

THERMODYNAMICS

OVERVIEW

Thermodynamics is the branch of chemistry that has to do with heat. One of the major topics of thermodynamics is the chemistry of spontaneity. A spontaneous reaction is a reaction that happens on its own without any outside influence. In this chapter we will be calculating three thermodynamic variables to determine the spontaneity of a reaction. Another important area of thermodynamics is calorimetry. In calorimetry we explore a concept known as specific heat and examine how it affects an object's ability to undergo a temperature change. By the end of this chapter you will be familiar with all of the different thermodynamic variables as well as the way in which to calculate each. You will also understand a bomb calorimeter, and see the types of questions that you will be expected to understand for the topic of calorimetry.

1. IMPORTANT THERMODYNAMIC VARIABLES

ΔH: Enthalpy (heat)

A positive ΔH indicates an endothermic reaction (heat absorbed)
A negative ΔH indicates an exothermic reaction (heat released)

ΔS: Entropy (disorder)

A positive ΔS indicates an increase in disorder
A negative ΔS indicates a decrease in disorder

ΔG: Gibbs Free Energy (spontaneity)

A positive ΔG indicates a non-spontaneous reaction
A negative ΔG indicates a spontaneous reaction

2. HOW TO CALCULATE THE IMPORTANT THERMODYNAMIC VARIABLES

When calculating thermodynamic variables, typically you first need to find ΔH and ΔS using their respective equations and values from a table, and then use those two variables to calculate ΔG. The following is explained below:

2a. Calculating ΔH:

$$\Delta H_r = [\sum (n \bullet \Delta H_{products})] - [\sum \{n \bullet \Delta H_{reactants})] \qquad \text{units: kJ}$$

(n = coefficient of that compound in the equation)
(ΔH_r = Enthalpy of the overall reaction)

Note: This equation looks very complicated; however, it is really easy. A lot of people get confused by the symbol "∑" which is a sigma. Those of you in sororities or fraternities might recognize this Greek letter. In math and science $\sum$ is used in equations to mean "the sum of". So this equation very simply reads "the sum of the ΔH values times the coefficients of the products, minus the sum of the ΔH values times the coefficients of the reactants."

When using this equation, you are usually given a balanced reaction and 4 out of the 5 variables in the equation. It is your responsibility to use the variables given and solve for the last variable. What I mean by this is that you are not always asked to calculate ΔH_r, so be prepared to be given a ΔH_r and have to solve for another variable in the equation.

Example: Calculate ΔH_r from the following information:

	CO +	$1/2\ O_2$	→	CO_2
ΔHf (kJ)	-100	0		-300

Step 1: Write equation for ΔH_r

$$\Delta H_r = [\sum(n \bullet \Delta H_{products})] - [\sum(n \bullet \Delta H_{reactants})]$$

Step 2: Plug in the given values:

$$\Delta H_r = [(1)(-300)] - [(1)(-100) + (1/2)(0)]$$

Step 3: Solve

$$\Delta H_r = -200 kJ$$

**Note: Whenever there is a single element in an equation, ΔH_f for it is always 0.

2b. Calculating ΔS:

$$\Delta S_r = [\sum(n \bullet \Delta S_{products})] - [\sum(n \bullet \Delta S_{reactants})]$$ units: J/K

Notice how the equation for ΔS is identical to the equation for ΔH. The only difference is that instead of plugging in ΔH values into the equation we are plugging in ΔS values. The other difference is the units are not in kJ but J/K

Example: Calculate ΔS_r from the following information:

	CO	+	$1/2\ O_2$	→	CO_2
ΔSf (J/K)	+195		+200		+215

Step 1: Write the equation for ΔS_r:

$$\Delta S_r = [\sum(n \bullet \Delta S_{products})] - [\sum(n \bullet \Delta S_{reactants})]$$

Step 2: Plug in the given values:

$$\Delta S_r = [(1)(215)] - [(1)(195) + (1/2)(200)]$$

Step 3: Solve

$$\Delta S_r = -80\ J/K$$

2c. Calculating ΔG:

Equation: $\Delta G = \Delta H - T\Delta S$ Units: kJ
(T = temperature in Kelvin)

This equation is called the Gibbs free energy equation. As I stated before, it is usually the last equation that we use when calculating each one of the variables. This equation is ultimately used to tell us whether a reaction is going to be spontaneous or not. If ΔG is negative, the reaction is spontaneous; but if ΔG is positive, it will be non-spontaneous. The only other thing to really watch for is to make sure the units of ΔS_r get converted to kJ/K first, since both ΔH and ΔG use units of kJ.

Example: Calculate ΔG from the given values for the following reaction at 25 °C and determine if it spontaneous or not:

	CO	+	$1/2\ O_2$	→	CO_2
Hf	-100		0		-300
Sf	195		200		215

Step 1: Write Gibbs free energy equation:

$\Delta G = \Delta H - T\Delta S$

Step 2: Convert the ΔS_r value previously calculated into kJ:

$$\Delta S_r = \frac{-80J}{K}\left(\frac{1\,kJ}{1000\,J}\right) = -0.08\frac{kJ}{K}$$

Step 3: Convert the temperature to Kelvin, since we must use the Kelvin scale:

T = 25 + 273 = 298

Step 4: Plug in all the values to Gibbs Free energy equation:

$G = -200 - [(298)\times(-0.08)] = -176.2$ kJ

Step 5: Determine Spontaneity:

The process is spontaneous because ΔG is negative.

2d. Determining conditions for spontaneity with the Gibbs Free Energy equation

Aside from calculating the value of Gibbs free energy, a lot of questions, particularly in the form of multiple choice, usually ask about spontaneity. So questions will usually ask about what specific conditions will lead to a spontaneous reaction and what conditions will lead to a non-spontaneous reaction. The way that you want to answer these questions is to use the Gibbs free energy equation and think about what will make the value of ΔG positive or negative (which determines if the reaction is non-spontaneous or spontaneous).

Example: "Under what conditions will a reaction *always* be spontaneous?"

Answer: An exothermic reaction that is increasing in entropy.

Explanation: An exothermic reaction has a negative ΔH. The Kelvin temperature is ALWAYS positive, and an increase in entropy has a positive ΔS. So $T\Delta S$ must be positive as well. Now, a negative number minus a positive number is always a negative number. This means an exothermic reaction that is increasing in entropy will always be a spontaneous reaction.

Other questions will involve cases with the same sign for both ΔH and ΔS, and the sign of ΔG will depend on the size of the temperature. For example, in the case of an endothermic reaction that is increasing in entropy, a positive number (ΔH) minus a positive number ($T\Delta S$) is only negative when the second number is bigger than the first. So in order for the reaction to occur spontaneously, it can only do so at a high temperature – to ensure the second number being bigger!

2e. Calculating Keq

Another crucially important reason for solving for ΔG is that it can be used to calculate K_{eq}. Calculating K_{eq} is typically a question that immediately follows the calculation of ΔG.

The equation for K_{eq} is:

$$K_{eq} = \exp\left(-\frac{\Delta G}{RT}\right)$$

Note that the "exp" refers to the e^x function. You can find this function on your calculator by hitting the 2nd function button and then the ln button. R is the gas constant with a value of 8.314 J/mol•K; and the temperature must be in Kelvin.

One thing to be careful about when using this equation is that the ΔG value must be converted to Joules since the R constant uses Joules. Usually ΔG is in kJ so don't forget to convert to Joules first!

Example: Determine Keq from ΔG = –176.2 kJ, calculated in the above example.

Step 1: Convert kJ to J:

$$-176.2\text{ kJ}\left(\frac{1000\text{ J}}{1\text{ kJ}}\right) = -176200\text{ J}$$

Step 2: Write K_{eq} equation:

$$K_{eq} = \exp\left(-\frac{\Delta G}{RT}\right)$$

Step 3: Plug in values:

$$K_{eq} = e\hat{}\left(-\frac{-176200}{(8.314)(298)}\right)$$

Step 4: Solve:

$K_{eq} = 7.69 \times 10^{30}$

3. HESS' LAW

Hess' law is an important method for determining the enthalpy of a reaction. Instead of being given all of the enthalpies for the products and the reactants in a reaction, you will be given a series of reactions with their respective enthalpies. The way to recognize a Hess's law question is when you are given some reactions with enthalpies and are asked to find an enthalpy for another reaction.

To solve a Hess's Law question we need to manipulate the reactions that we are given in order to have them sum to the reaction that we are asked for. What I mean by this is that you need to go one compound at a time in the reaction that you are solving for and look to see where that compound is present in the series of reactions that you are given. If we need a certain coefficient in front of the compound in the final equation, then we multiply the initial reaction by a number to generate that coefficient. If the compound must end up on the opposite side of the final equation than the side it starts on. We must invert the entire initial reaction is on the wrong side of the equation, you must flip the reaction.

Now, the crucial point is that whatever it is that you need to do to each initial equation, you must also manipulate the enthalpy for that specific reaction in exactly the same way. So f you multiply a reaction by 2, you must multiply the associated enthalpy by 2. If you flip the reaction, you must negate the enthalpy. Once you are done with your manipulations, you simply add all of the enthalpies together and that is your enthalpy for the reaction you were looking for.

Example: Find ΔH_{rxn} for the following reaction: $C_3H_8 \rightarrow 3\ C + 4\ H_2$ using:

$CO_2 \rightarrow C + O_2$	$\Delta H = 390$ kJ	(1)
$2\ H_2O \rightarrow 2\ H_2 + O_2$	$\Delta H = 500$ kJ	(2)
$3\ CO_2 + 4\ H_2O \rightarrow C_3H_8 + 5\ O_2$	$\Delta H = 2000$ kJ	(3)

Step 1: Start with the first compound C_3H_8 in the reaction that we are solving for. Notice it is present there as a reactant but in reaction (3) it is a product. So we have to invert (3), and then also negate the ΔH value for reaction (3):

$C_3H_8 + 5\ O_2 \rightarrow 3\ CO_2 + 4\ H_2O$ $\Delta H = -2000$ kJ

Step 2: The next compound is C. Carbon is present reaction (1) as a product. The only thing that we need to manipulate is the coefficient in front of the carbon. In the reaction that we are solving for C has a coefficient of 3; however, there is a coefficient of 1 for C in reaction (1). Therefore, we must multiply all of (1) by 3. By multiplying the reaction by 3, we must also multiply the ΔH value by 3:

$3\ CO_2 \rightarrow 3\ C + 3\ O_2$ $\Delta H = 3 \times 390 = 1170$ kJ

Step 3: Lastly we focus on H_2. It is located as a product in reaction (2). In the reaction that we are solving for it has a coefficient of 4 and in reaction (2) it has a coefficient of 2. So we multiply the second reaction (2) AND the enthalpy by 2:

$4\ H_2O \rightarrow 4\ H_2 + 2\ O_2$ $\Delta H = 2 \times 500 = 1000$ kJ

Step 4: Now add all of the reactions and the enthalpies to get the final answer:

$C_3H_8 + 5\ O_2 \rightarrow 3\ CO_2 + 4\ H_2O$	$\Delta H = -2000$ kJ
$3\ CO_2 \rightarrow 3\ C + 3O_2$	$\Delta H = 1170$ kJ
$4\ H_2O \rightarrow 4\ H_2 + 2\ O_2$	$\Delta H = 1000$ kJ
$C_3H_8 \rightarrow 3\ C + 4\ H_2$	$\Delta H = 170$ kJ

*Note that when the three manipulated reactions are added, certain reactants cancel with some products to leave us with the reaction that we were looking for.

4. CALORIMETRY

Calorimetry is the next important topic of thermodynamics. The questions involving calorimetry involve something known as a bomb calorimeter. A bomb calorimeter is a device that is used to insulate a chemical reaction. In lab, you will usually use a coffee cup to act as a bomb calorimeter. The reason why we use a Styrofoam cup to drink hot drinks is because it insulates the heat and prevents us from burning our hands. Just as Styrofoam insulates the heat when we are drinking, it also insulates it in a chemical reaction, so can serve as a calorimeter.

When using a calorimeter, we can assume that all of the heat within the system stays in the calorimeter and is not lost to the environment. With this assumption we can go ahead and say whatever heat is lost by one object must be equal to the amount of heat that is gained by the other. The usual way that we see this presented is in a case where hot metal is dumped into a Styrofoam cup of cold water, as in the following illustration:

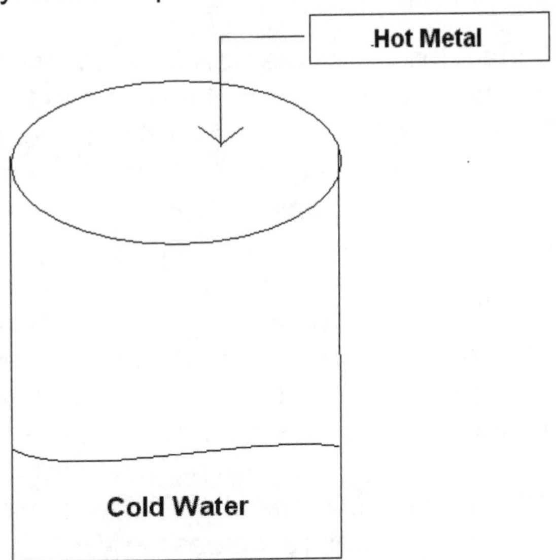

Since heat lost by the metal must equal to heat gained by the water we can say:

$q_{water} = -q_{metal}$ (where q = heat)

Common sense tells us that the hot metal will get cooler and the cold water will get warmer since we are dumping something hot into it. However, at what temperature is the overall process going to equilibrate? Will the temperature stop closer to that of the original temperature of the metal or of the water?

This question has to do with a concept known as **specific heat**. Specific Heat tells us how easy it is for an object is to undergo a temperature change. Different metals have different specific heats; resulting in some metals becoming colder quicker than other metals.

The way that we solve for specific heat is the following equation:

$q = mC\Delta T$ (I remember this as looking the medical admission test MCAT)

Components: m = mass in grams
C = Specific heat (in J/g•°C)
ΔT= Change in temperature

Next, plugging in the equation for q to the above equation of $q_{water} = -q_{metal}$ we form a new equation:

$mC\Delta T_{water} = -mC\Delta T_{metal}$

We can use the above equation in all bomb calorimeter problems to obtain the specific heat of an unknown metal. Typically, we will be given 5 out of the 6 variables and asked to solve for the last variable.

There are a few important things to note about the equation. The first important thing to do is to keep all of the variables organized. Make sure you realize that all of the variables on the left side of the equation are for the properties of the water, and all of the variable on the right side of the equation are for the properties of the metal. Second, you should remember the specific heat of water as you may not be given the value in the problem. The specific heat of water is 4.18 J/g•°C
Lastly, it is important to know that ΔT means $T_{final} - T_{initial}$. This will be particularly important if you are ever asked to solve for the initial or final temperature. In a case where you are asked to solve for the final temperature when given an initial temperature of 25° C you need to use an expression for ΔT that looks like (x-25) for example. In this case you would solve for x to obtain the final temperature.

Example: When 30 g of an unknown metal at 200° C is dropped into 150 mL of water initially at 25° C, the water temperature increases by 2.0° C. What is the specific heat of the unknown metal?

Step 1: Write the equation for calorimetry

$mC\Delta T_{water} = -mC\Delta T_{metal}$

Step 2: Organize all of the variables

Water	Metal
m=150	m=30
c=4.18	c=?
$\Delta T=\|(27)-(25)\|=2$	$\Delta T=\|(27)-(200)\|=173$

**Note that the volume of water in mL is equivalent to the mass of the water in grams because 1mL = 1g of water.

Step 3: Plug the values into equation:

$$(150)(4.18)(2) = (30)(C)(173)$$

Step 4: Solve

$$C = 0.24\ J/g{\cdot}°C$$

**Note that by being given the increase in temperature of the water we were able to calculate the final temperature to be 27 which was used to find the change in temperature of both the water and the metal.

SUMMARY

In this chapter we addressed important areas of thermodynamics. I strongly recommend that you become familiar with the thermodynamic variables that were explained in this chapter in terms of what they represent and what is meant when the sign is positive and negative. Be sure you know how to solve for each variable and be aware of the units for each. These thermodynamic variables usually constitute a large portion of test questions. Also make sure that you recognize when to use Hess' Law and understand how to manipulate both the reactions and the enthalpy values. For calorimetry, be aware of why a bomb calorimeter is used and how it allows us to solve for specific heats of unknown metals.